I0436731

THE ART AND SCIENCE OF SECURITY

Practical Security Applications for Team Leaders and Managers

Joel Jesus M. Supan

Order this book online at www.trafford.com
or email orders@trafford.com

Most Trafford titles are also available at major online book retailers.

© Copyright 2012 Joel Jesus M. Supan.
All rights reserved. No part of this publication may be reproduced, stored in a retrieval
system, or transmitted, in any form or by any means, electronic, mechanical, photocopying,
recording, or otherwise, without the written prior permission of the author.

Printed in the United States of America.

ISBN: 978-1-4269-8204-0 (sc)
ISBN: 978-1-4269-8205-7 (hc)
ISBN: 978-1-4269-8207-1 (e)

Library of Congress Control Number: 2011913321

Trafford rev. 08/15/2012

 www.trafford.com

North America & international
toll-free: 1 888 232 4444 (USA & Canada)
phone: 250 383 6864 ♦ fax: 812 355 4082

To my wife, Susan and sons Martin Thomas, Stephen Isaac and Jan Nathaniel, who all provided the inspiration for this book and who allowed me the space to grow into what I am now.

To my company, which provided the fertile ground on which the concepts were affirmed, nurtured and put to good use.

PREFACE

Security and safety are at the second level of the hierarchy of human needs according to Abraham Maslow. Both are on top of the physiological needs, which include air, food, shelter, excretion among others. The need for security and safety encompasses the body, health, family, resources and property.

The need and means for Safety and Security may be instinctive but the varying hazards that threaten these resources are so numerous that instinct alone is not sufficient to satisfy those needs. In time, man's experiences have taught him to develop various means of security for himself and his resources.

The types of hazards that confront man differ depending on the circumstance he is in. Although the hazards may differ, the varying means of security are all based on common principles. They require the same aspects or means for security to be achieved.

After the need for safety and security is satisfied man develops the need for social relationship and to belong. These needs do not stop with simple relationships. One needs also to belong to a more structured organization such as a team. During this stage, he either becomes a member or the leader of a team.

The Team Leader is responsible for the utilization, preservation and protection of the company and its resources. These should be done because resources are necessary for the team to achieve its objectives.

One of the timeless and basic principles of leadership is, "A leader must be technically and operationally more proficient than his members."

The foregoing are the premises by which this book was written. It is intended to make leaders understand the real meaning of security and realize why this is one of their primary responsibilities.

This book aims to provide practical guidelines for the application of the principles and aspects of security in the daily performance of their roles in their organization, their families and their personal lives.

Security transcends the general domain of guards, law enforcement agencies and the military, the law enforcer and the military man and the suppliers of security hardware.

This book will also demonstrate the limitations of the generally accepted means of security such as guard systems and pure technology if they were to be used as stand-alone systems.

Moreover, it aims to provide the bridge if not fill the gaps between other fields such as Business and Competitive Intelligence, Risk Management, Insurance and Business Continuity and integrate them to one whole body of discipline. These concepts and practices, often mistakenly believed as not within the domain of security, are actual functions of security.

Teams are created, organized or simply become as such because of an idea and an objective. The team may be as small as the basic two-man team, a large organization with thousands of members, a nation with a million members or the world with billions of members.

A team may refer to an individual or a family with an objective of bringing up productive children and having happiness and contentment in one's home. It may refer to team that runs a corner store or a small unit, a department or a division of a large organization. It may be the organization itself. Each smaller unit within the organization has the objective of supporting the overall objective of a larger unit. A team can be the members of a community or the citizens of a country.

The one who leads the team to its objective is the Team Leader. To reach the team's objective, it needs resources that the members must protect from loss or damage.

Security requires the presence of both the feeling of security and the realization of the set objectives.

These requirements are the basis for the expansion of the context of security from the conventional to the common everyday activities by the man on the street and the Team Leader.

A closer look by a learned security practitioner on how both systems as applied, however, would show that they are more reactive than preventive.

As a case in point, guard training gives emphasis on law enforcement, appearance, courtesy, weapons training, unarmed self-defense, fire-fighting, public relations and drills; while technology gives emphasis on deterrence, detection and documentation.

All these provide the consumers of security services a feeling that their facilities or resources are secured. This is because they do not have formal benchmarks upon which they can gauge the effectiveness of their security system, except when they are able to catch an offender after damage has been done. There is no way to measure if the expenses for their security system were well worth it.

The cost of security presents another issue altogether. College courses as well as post-graduate courses in business and management do not offer subjects on security. Yet, any well-meaning entrepreneur knows that security is always one of the biggest items of expense in their budget. Others, on one hand, are just too small to spend so much money as to protect their businesses. What most people don't realize is that their security lies on a clear understanding of the application of security principles and its aspects. On the other hand, other companies hardly spend for their security and recognize its value only after they have incurred considerable damage.

The mission of a Company Security Department is commonly relegated to providing physical security, guard force management and supervision and investigation.

It is upon this backdrop that a security model has been devised, the Stonewall Security Model. The model is named after the company that created it.

While this book covers all the aspects of security based on the Security Model, emphasis on execution shall be limited to the aspects that are conventionally performed by security team leaders. These are the Personnel, Information, Operation, and Physical Aspects of security. The Environment and Reputation Aspects of security shall be discussed briefly in separate chapters.

The applications of these aspects of security have given way to the idea for the title of this book, "The Art and Science of Security." Security as a science involves time tested principles, logic and good reasoning. It requires the use of technology and the laws of physical sciences. On the other hand, Security as an art involves the use of various disciplines and skills that go with the discipline. It requires the use of common sense in creating ideas and concretizing those ideas with propriety, correctness and timeliness to address the ever-changing conditions in the environment.

As previously stated, this book is intended to teach and guide Team Leaders on how to preserve and protect the resources of the team so it can achieve its objectives. The importance for a leader to be more proficient than his members cannot be overemphasized. It's important that he knows when his members are doing their jobs right and for him to know how to correct them if they don't. As one great leader said, "I just won't order my people to do things, which I cannot and won't do."

Security should be the second language of a leader. He should be able to share to his members the concept that security is everybody's concern and that it is a shared responsibility for them to develop a culture of security.

Lastly, there is one aspect of security that is apparently omitted. But in reality, it is not. This is Spiritual Security. Spiritual Security is the aspect of security that reinforces all other aspects of security. No undertaking or goal can be achieved without this aspect. The other aspects discussed in detail in this book are all based on reason. But Spiritual Security need not be explained as it is based on faith. It has been the guiding force behind the writing of this book. No one can have a better sense of security than by having faith and believing that the ultimate security is having peace in the great beyond after all the material things we protect on earth have come to pass.

Once, one of the directors of our organization asked me, "Could you still sleep soundly at night with all the concerns you have on the risks that threaten our organization?" Our organization is a transport company that conveys millions of people each year. It is most susceptible to terrorism and regularly receives terrorist threats. Regardless of their being hoaxes or not, no officer slept until my team had done due diligence and agreed that the threatening call was not real. Going back to the question, I answered, "Yes, I could. That's because, I say a little prayer to Him and ask help for our protection before I go to bed each night." To which the director replied, "I am glad that someone is taking care of the more important matters in running the business."

TABLE OF CONTENTS

Preface

SECTION I

The Fundamentals and Basic Concepts of Security

A Team Leader and any student or practitioner of security must have a clear understanding of its fundamental principles before attempting to practice any of its specialized aspects. It is in the mastery of these fundamentals that will make a student learn the rationale behind all security measures and practices. These fundamentals are the foundation for the security profession to become an institution. It is with this institution that security will be recognized as a profession.

Since this book is pushing for security integration even for those who are not practitioners such as those team leaders of functional units of a company, knowledge of security is a must.

Learning the fundamental concepts and principles stated in this section is essential in the formulation of security programs and policies, which will be described in detail in the latter sections of this book.

CHAPTER 1

Basic Security Principles and Concepts

WHAT IS SECURITY?

Security is a word used in a lot of ways and applied to a lot of things. To a bank, it is the collateral required from borrowers before they can "secure" a loan. To a country or state, it refers to the economic, social and political stability and the freedom from foreign intervention. To states within a geographical region or states with common interests, it means a sound and stable coexistence or relationship with each other. To a public enterprise, it refers to shares, bond, stocks and documents that give a right to an individual to own properties.

To most people, the meaning of security is the provision for physical protection such as policemen, military men, guards, watch dogs, fences, walls, alarms, lights or cameras. Security also implies safety as it can also mean a state of freedom from harm, damage or loss. It is a means to protect lives, properties, tangible or intangible assets, such as reputations, and material resources needed to pursue one's goals.

Security is a means from where "sense of security" is obtained. Security is implied in one way or the other in other organizational undertakings and concepts such as risk management, loss prevention and control, business continuity, corporate good governance and business intelligence.

In its broadest sense, security provides a predictable environment where one can pursue his objectives without fear from the occurrence of mishaps and their effects when they occur.

Under any condition, security can protect a person against any hazard and at the same time provide a sense of security. Feeling of security alone does not guarantee protection. In the same light, the value of real protection could not be appreciated if one does not have peace of mind or a sense of security.

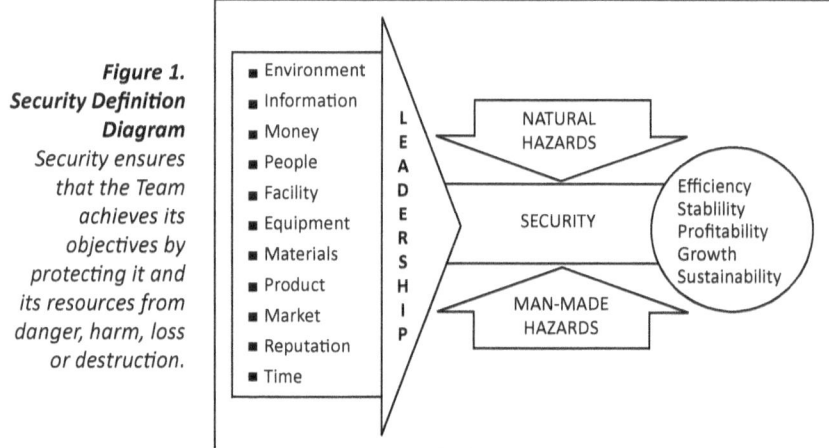

Figure 1.
Security Definition
Diagram
Security ensures that the Team achieves its objectives by protecting it and its resources from danger, harm, loss or destruction.

BASIC OBJECTIVES OF AN ORGANIZATION

The pursuit of security must always be aligned with the objectives of a team such as in an organization or a business enterprise, a household, an individual or any undertaking.

All teams, big or small, have common objectives regardless of how these are expressed. Companies encapsulate these in their mission and vision statements. A family describes it as a happy home. An individual simply calls it a dream or an ambition.

It is common for organizations to come up with fancy–worded Corporate Mission and Vision Statements. Often, these statements are so verbose and long-winded that no one in the organization could express it verbatim after even a short while. A saying goes that "if you can't express it you can't do it." Thus, if one cannot express his objectives, he cannot develop the mind-set that will be the guiding force to achieve his purpose.

The simpler the objectives are stated, the easier for the members to understand, memorize, internalize, express and perform to achieve them.

Take a closer look at some mission and vision statements of organizations, their reason for being is punctuated by a common denominator expressed in two simple words, "Make Money."

And for even bigger business organizations, their mission can be said in four words: "Make Much More Money." Simply put, one's "Mission and Vision" cannot be achieved without "Money."

Be that as it may, the organization's mission and vision are posted on office walls as a ready reference and reminder to team members. The team leader must ensure though that these objectives are retained in their hearts.

A discussion on Personnel Security in Chapter 3 mentions one Security Principle", *a leader must keep his team members informed.*" More than that, he must also ensure that his members know the task they need to perform to achieve their mission. Thus, the simpler the mission is stated the easier it will be understood. It is in the understanding of the objectives that the team can focus on the security actions needed to achieve it.

Most team objectives can be summarized in five simple words: Efficiency, Stability, Profitable, Growth and Sustainability.

These objectives apply not only to business organizations but also to the individual, the family, and even a country or any large entity. They form the hierarchy of organizational objectives. This means that the organization must achieve efficiency first before it can have stability. It also means that once the higher objectives have been achieved, it is essential to maintain the lower objectives.

These objectives are further explained as follows:

1. **Efficiency**

Efficiency is simply defined as making the most of ones resources. All companies, just like everyone else, endeavor to make the most out of their resources or investments. They have procedures written to ensure nothing is wasted. Efficiency is measured in terms of productivity, where the output should have more value than the input in the short or long term.

The same definition is applied on households and individuals. To the family, the home is their refuge and there are unwritten house rules that are aimed at conserving resources, which they need to be able to rest after a day's work.

The same could also be said of individuals who have set their goals in life. They endeavor to conserve and stretch their personal resources to achieve their goals.

2. **Stability**

Stability is a team's ability to withstand the effects of external conditions. While efficiency involves the control of internal conditions, stability is the resiliency of the organization against threat from outside of the organization. Examples of external interventions are the adverse effects of a bad economy or the disaster due to natural causes.

To a family, this objective is about their capability of staying and living together by supporting each other and bonding together thru love, caring and nurturing, regardless of the trials that come their way such as sickness, disasters or loss of job by the bread winner.

To an individual, the attainment of this objective implies availability of support that can make one withstand pressures and challenges.

3. Profitability

Profitability is achieved when the value of the team's output is greater than its input. It is the main reason for anyone who goes into business in the first place.

To a household or an individual, the attainment of this objective means having amassed disposable income or resource in excess of its basic needs.

To an individual, this is the savings from his income that will allow him freedom to enjoy excesses.

4. Growth

A team must grow to be able to deliver the needs of an expanding market brought about by the rise in population and in the economy. A team that does not grow will eventually die.

Companies also need to grow to give room for their employees' professional and economic growth.

This is the reason why once a team becomes profitable it endeavors to expand its business.

As its employees advance in their respective professions, so are the needs of their families. Their family members grow in number. They need to sustain their resources to support their families.

To an individual, this means the acquisition of more personal assets and social mobility and widens his areas for opportunities.

5. **Sustainability**

Being sustainable is achieved when an organization is able to perpetuate its existence. Most people think that owners of companies want their business to grow because they want to earn more money. But to a certain extent, more often than not, they are obliged to sustain the livelihood of those who work for them and then expand some more to extend opportunity of livelihood for others. This is one way to give back to the society that supported them. Upon reaching this objective business and the society becomes symbiotic.

Being sustainable to a family starts when children become responsible and productive members of society. By growing responsible and productive children, the family contributes significantly to the society that they are a part of.

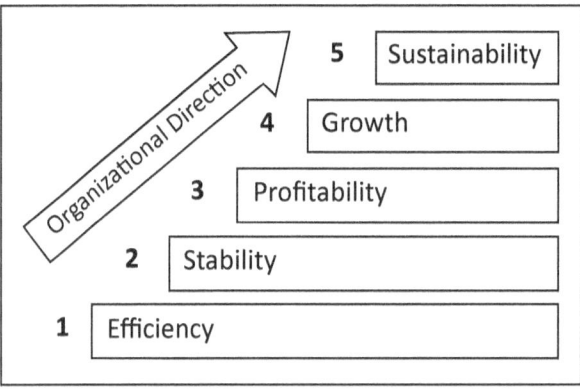

Figure 2. Hierarchy of Organizational Objectives.

The lower objectives must be achieved before the next higher objective is achieved. The basic and lower Corporate objectives are maintained even after the higher objectives were achieved.

It is upon these five objectives that all security actions like all the other functions and undertakings of the teams will be focused.

THE BASIC RESOURCES OF AN ORGANIZATION

The resources of the team are the means by which it can achieve its objectives. These resources, together with the hazards and risks that threaten it are the objects of security.

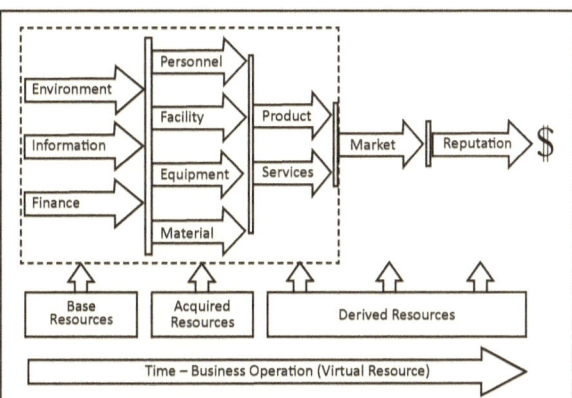

Figure 3. Classification of Organizational Resources as to Origin.

Organizational Resources can be classified as to origin as Base Resource, Acquired Resource, Derived Resource and Virtual Resource.

Resources are the items that need to be preserved, conserved and protected from loss, wastage or destruction. It is in securing, protecting and conserving these resources against damage or loss that a team can achieve its objectives. The Team Leader is responsible for the conservation, preservation or proper utilization of these resources. Resources are classified generally into the following:

1. Environment
2. Information
3. Finance
4. Personnel
5. Facility
6. Equipment
7. Material
8. Product
9. Market
10. Reputation
11. Time

The detailed descriptions of these resources are as follows:

1. **Environment**

Environment is a resource that refers to the natural, physical, social and political and the industry that surround the organization. This is classified as a base resource as it provides the opportunity and inspiration for the creation of an organization. The environment indicates that there is a need that must be satisfied which the team will ultimately try to capitalize on by creating a service or product.

This is the resource that sustains the existence of the

organization. It also refers to the natural resources from which materials used to produce the product are taken from.

There are specific assets that constitute the elements of this resource. For example, natural assets refer to land, sea, atmosphere, even bodies of water. To be more specific, a river is the principal asset of a hydro plant; just like an arable land is the primary resource of an agricultural company.

The physical environment refers to support structures man-made or natural in and around the facility. Examples are roads, power lines, buildings and grounds. Good road facilitate the easy conveyance of materials, products and people.

Social environment includes neighbors, the community and the society that support and serve the organization by providing services or by their patronage.

Political environment is the resource that refers the different government functionaries. It includes the people managing the affairs of these government offices. These are the people who administer the laws and regulations that have conferred to the team or organization its legal existence. They provide the services, polices and guidelines for the equitable utilization of resources so as to create an atmosphere for the fair exchange of goods and services.

Lastly, Industry Environment refers to the marketplace, the players and competitors within the sector of endeavor in which the organization belongs. Each of the competitors in the industry constantly wants to get the biggest share of the pie that represents the market and the resources by any means. Alliances or cutthroat competitors are formed, developed or disbanded with regularity.

To a family, its environment include a clean and orderly surrounding, kindred relationship with his neighborhood and community and his obedience to the laws of the country and the regulations that govern his community.

2. Information

Information is a resource that refers to information or data about employees, policies, plans, projects, designs, patterns, business application software, production, sales and other business transaction data, digital or scripted, documents describing systems, procedures, rules and regulations, company intelligence of the environment such as political, social, economic and industry, legal cases and their proceedings and rulings, and even organizational and individual failures, idiosyncrasies, misdeeds and mishaps.

Information is the product of an idea created or perceived by human intelligence and imagination that is expressed in writing or orally. It is transmitted from the creator to another person who has the skill to concretize these ideas into products or services. Information in turn, is converted to more valuable commodities for one to realize his growth.

Information is the lifeblood of the organization. As such, if it were lost, disclosed to the competitor or distorted, the team would find it difficult, if not fail, to achieve its goals. Like blood, if it were drained or contaminated, the body that bears it dies.

To a household, information means the assets of the family or any derogatory information that the family keeps.

To an individual, information refers to his knowledge obtained from education and experience as well as the state of his physical, economic and social condition.

There are two types of information that an organization must protect. These are the good and the bad information. If good information were lost or distorted, the organization would suffer delay or considerable, if not, total loss. If it were disclosed, the competitors would take advantage of, compete with the organization and share with its goals. Bad information on the other hand, if disclosed to the competitor, could be used against the organization that would challenge its good reputation.

JOEL JESUS M. SUPAN

3. **Finance**

>   Financial resources include all monies, drafts, stock certificates, letters of credit and any negotiable instruments. These assets are necessary to sustain the day-to-day business transactions of the organization. Money is the universally accepted medium of exchange used to acquire goods and services necessary in producing products or services. Need for money is common to all teams, companies, households and individuals.

>   Without money, the machine that is the organization will not run. Money, being a commodity for convenience, is susceptible to human malfeasance and misfeasance.

4. **Personnel or Human Resource**

>   Personnel are the resource, which refers to all regular employees, contractual employees, and employees of contractors. The prevailing predisposition of business organizations on contractual employees is that they are not employees of the company so that they are not considered as assets. This is on account of legalities of employer-employee relationship espoused by labor laws. But, as a matter of principle and purpose, all persons providing direct services to deliver the needs of the organization are employees.

>   Services of contractual employees and employees of service contractors are acquired by the organization to provide necessary labor, which the organization cannot do without. In this light, they are assets of the company. On the other hand, to deliver their required services, contractual employees have access to all the other company resources and facilities. In this regard, they can also become hazards to the organization like all regular employees.

>   Human resource is likened to the engine of the vehicle. The team members, like the employees of a company, are the prime movers of the organization. It is the employees who create ideas, design systems to provide frameworks for those ideas, make

decisions, execute and work with their hands to concretize those ideas into products or services.

To the household, human resource refers to the household members such as the parents, children, house helpers, drivers, gardeners and baby sitters. Every member of the household has a specific obligation and task to perform to maintain and keep the home a safe refuge.

To the individual, the human resource is himself. He has his human and his human faculties of intelligence, reasoning, five senses and his physical abilities to perform work.

5. Facility

Facility is a resource that refers to the real assets, buildings and other structures that house the offices or production equipment, where the essential activities of the organization are performed. They could be owned or leased. They are generally susceptible to natural or social phenomenon.

To the family, facility is their house and other structures which protect the family from the elements.

To an individual, these are his personal room, offices or other physical resources where he practices his trade.

6. Equipment

Equipment and machineries are resources which are used as implements to produce or enhance the productivity of the members of the organization or the maintenance of other facilities and equipment. These are the tools, which aid the personnel in creating a product or in rendering services economically, accurately, consistently and expediently. The resources that belong to this classification are the machinery, communication equipment, vehicles, power tools and power supplies. Conveyances and high technology hardware that produce and store data are also included in this classification. These resources are generally susceptible to destruction or obsolescence due to human design, inadvertence, ineptitude or indifference.

To a family, this refers to the appliances and implements for convenience in doing house chores such as ovens, washing machines, refrigerators, vacuum cleaners and flat irons.

To an individual, these are the tools of his trade such as personal vehicles, laptops and cellular phones.

7. **Materials**

Materials are the consumable resources which the team needs to produce its product or services. They may also be used for administrative or production support. Examples of assets belonging to this classification are office supplies and other materials used for the productions of the end products or the delivery of services such as raw materials, paper, ink stapler, utensils, etc. Loss of these assets would impede the critical activities of the organization. Being portable, salable or consumable, they are also very susceptible to human inadvertence, ineptitude, indifference or malfeasance.

8. **Products and Services**

Products and Services are the outputs which are bought or availed of by customers. When sold, they represent the return of the team's investment. These assets, aside from their materiality, bear the name or brand of the organization. Products and services are the commodities that are directly exchanged for the revenue, which the organization aims to achieve at the outset. This is the statement of one's contribution for the good of the buying public. This is the commodity that defines the identity of the team and its acceptance by the consumers that will sustain its existence.

To a household, this refers to the comfort, peace and quiet, happiness and contentment of its members.

To an individual, this is his personal talent or the output of such talent that would satisfy the needs of others.

These resources are likewise very susceptible to human inadvertence, ineptitude, indifference or malfeasance. They can also be affected by natural and social phenomenon.

9. **Market**

Market is the resource that potentially and actually buys the team's products and services. It is the resource that values and puts value into a company's products or services. It can be an individual or institutional buyer, a current user or a potential user of the products or services produced and offered by the organization.

The market is the direct source of the revenues aimed for by the organization.

To a household, market refers to the household members themselves, as they are the consumers of the products of the home, the product being "peace of mind".

To an individual, his employer and his coworkers are his market for his talent, skills and services.

The buyers of the organization's products or services could be lost to the competitors if the products or services fail to meet their expectations.

The market could also be lost due to economic conditions which could be adversely affected by environmental, social and political phenomenon.

10. **Reputation**

Reputation is a resource that puts an organization, its name and brand at a high esteem by the market and the industry. This is marked by the acceptance of the consumers of the team's services or products. It is also the recognition of the team's value compared to the rest of the industry by its suppliers and stakeholders.

When combined with the quality product and the mutual trust between the consumer and the team, it defines a more valuable resource than any other individual resource. This is what creates and sustains the team's brand.

Reputation does not only refer to the reputation of the organization as a whole but extends to its employees, products or services and brand that represents it.

To a household, it is the kind esteem and high regard that the community gives the family that lives in it.

To an individual, his reputation is the asset that would sustain his good relationship with those he deals with.

It is these institutional, individual and brand reputations that sustain and expand the market and ultimately make the organization achieve all its set objectives.

Reputation is susceptible to human inadvertence, ineptitude, indifference or malfeasance.

11. Time

Time is the resource that provides the space and opportunity for the team and its personnel to perform their function in complete synergy. While it may seem to be inexhaustible, time spent can no longer be recovered and any loss of it is without restitution.

Time with family members is investments as it provides opportunity to mold children to have strength of character.

Time with others provides the opportunity to strengthen his relationship to the community and the society which can provide growth and morale support when needed.

Time with one's self provides for quiet reflection or simple relaxation.

Figure 4. Table of Comparison of Objectives and Resources of Small Organizations. *Organizations, regardless of their sizes have the same objectives. They need the same resources to achieve their objectives.*

Organization	Individual	Family	Corner Store	Large Company	Community	Country
Objective	Success, Happiness	Happiness, Contentment	Money	Money	Peace and Order	Gross National Product, Happiness
Management	Self	Parents	Owner	Management Team	Community Leaders	Government
Resource Classification	**Examples of Resources**					
Environment	Profession, Friends, Market	Neighborhood	Neighbors, Households	Market Conditions, Location	Neighborhood	Land, Seas, Neighbors
Information	Knowledge	Information on members	Price, Offers	Business Plans, Trade Secret	Information on Residents	Intelligence
Finance	Salary, Savings	Family, Income	Revenues	Revenue	Revenue	Revenue
Personnel	Self	Household, Members	Store Personnel	Team Members	Residents	Citizens
Facility	House, Office	House	Store	Office, Building	Schools, Markets, Roads, Hospitals	Infrastructure
Equipment	Cell Phone, Laptop	Appliances, Car, Furniture	Coolers, Carts, Cash Register	Machinery, Equipment	Street Lighting, Transportation	Industry, Utilities
Material	Personal Belongings	Foods, Clothes	Paper bags, Boxes	Raw Materials	Foods, Clothes	Natural Resources
Product or Services	Labor, Ideas, Responsibilities	Food, Shelter, Rest, Recreation	Merchandise	Product, Services	Services, Peace and Order	Socio-Economic - Political Stability
Market	Employer, Clients	Family Members	Customers	Clients, Patrons	Residents	Citizens
Reputation	Reputation	Reputation, Good Name	Reputation	Reputation	Reputation	Reputation
Time	Time	Time	Time	Time	Time	Time

Time as a resource is deemed as a virtual resource. While it is not tangible, it is no less real. It is susceptible to loss by the undoing of any of the threats caused by nature or man.

In the practice of security, it is essential that the Team Leader must recognize the importance of these resources individually or in combination with the others. Each of these resources may have varying degrees of importance or criticality to the team. The damage or loss of any of these resources can be detrimental to the team.

Equally important is the Team Leader's knowledge of the vulnerability of these resources against specific hazards. The study of asset criticality or impact and vulnerability shall be provided in detail in Section III of this book.

HAZARDS AND THREATS:
THE ENEMIES OF THE ORGANIZATION

A hazard is a situation or a condition, tangible or intangible that poses a threat to life, health, property, environment or one's reputation. Hazards are the inherent enemies of any organization. As such, learning about them is one of the fundamental requirements in the study of security. In the same way, knowing one's enemy is a fundamental requirement to win battles and wars. For a team to attain its objectives it must know and understand by heart the hazards and dangers that threaten to destroy or erode its resources.

MODES OF HAZARDS

There are different modes to which the threats of hazards belong and are defined. These modes apply to all types of hazards.

A hazard in a **dormant mode** is when it merely has the potential to be hazardous. No person property or environment is affected by this condition. A volcano in an isolated and uninhabited island is an example. Should the volcano erupt, there are no people and properties that could be affected due to its location.

A hazard is on a **potential mode** when it is in the position or a situation to harm persons or destroy properties or environment. Hazards in this mode need to be assessed to determine the risks it could bring.

A hazard in an **active mode** when it is certain to cause harm as no intervention can be made before the incident.

If a potential hazard has been identified and interventions have been made to ensure it does not progress to become an incident, the hazard is deemed to be mitigated. The interventions may not guarantee that there would no longer be a risk, but it is likely to have significantly reduced its potential.

Hazards may vary according to the locality, type and nature of business operations and the degree of their impact.

CLASSIFICATION OF HAZARDS

Hazards are basically divided into three major categories as to their cause. They are natural, man-made and activity-related hazards.

1. **Natural Hazards**

These hazards are generally caused by natural phenomenon. The following are examples of natural hazards:

    1.1   Drought
    1.2   Earthquake
    1.3   Epidemic or Pandemic
    1.4   Flood
    1.5   Heat Wave
    1.6   Hurricane or Typhoon
    1.7   Infestation
    1.8   Landslide
    1.9   Lahar

1.10 La Niña

1.11 Lightning

1.12 Pestilence

1.13 Plague

1.14 Storm

1.15 Tornado

1.16 Tsunami

1.17 Volcanic Eruption

2. **Man-Made Hazards**

These hazards are acts of omission or commission of man. Acts of omission are caused by one's ignorance, indifference or negligence. Acts of commission are actions that are motivated by either need or greed.

Man-made hazards are further classified by the objects of the act of man. They are as follows:

2.1   Man-made Hazard Against Properties

2.1.1   Arson

2.1.2   Intrusion

2.1.3   Sabotage

2.1.4   Squatting

2.1.5   Theft

2.1.6   Vandalism

2.2   Man-made Hazards Against or By Oneself

2.2.1   Accidents

2.2.2   Absence from work

2.2.3   Drug Addiction

2.2.4   Ignorance or Ineptitude

2.2.5   Negligence

2.2.6   Violation of Regulation

2.3 Man-made Hazards Against Persons or Criminal Acts

2.3.1 Harassment

2.3.2 Grave Threat

2.3.3 Fist fight

2.3.4 Assault

2.3.5 Robbery

2.3.6 Kidnapping

2.3.7 Homicide

2.3.8 Murder

2.4 White-Collar Crimes Against One's Organization

2.4.1 Tardiness

2.4.2 Fraud

2.4.3 Theft of Information

2.4.4 Piracy

2.4.5 Discrimination

2.4.6 Disclosure of Information

2.4.7 Abuse of Time

2.4.8 Abuse of Authority

2.4.9 Abuse of Privilege

2.4.10 Harassment

2.4.11 Violation of Policies

2.4.12 Non-Compliance to Procedures

2.4.13 Omission of Duties

2.5 Social or Political Hazards

2.5.1 Activism

2.5.2 Dissidence Riot

2.5.3 Labor Unrest

2.5.4 Rebellion

2.5.5 Strikes

2.5.6 Terrorism

2.5.7 War

3. **Activity–Related Hazards**

Activity–related hazards are achieved by the undertaking of a certain activity or condition. This will be eliminated by stopping the activity, the correction of condition or removal of the cause of a hazardous condition. They are generally implied as safety hazards.

3.1   Hazardous Activities

3.1.1   Flying
3.1.2   Skydiving
3.1.3   Spelunking
3.1.4   Cliff Climbing
3.1.5   Combat Sports
3.1.6   Contact Sports
3.1.7   Other hazardous jobs

3.2   Safety Hazards within Ones Surroundings

3.2.1   Chemical Spill
3.2.2   Defective tools, equipment and materials
3.2.3   Electricity
3.2.4   Excessive noise
3.2.5   Flammable materials
3.2.6   Improper stacking
3.2.7   Inadequate guards or barriers
3.2.8   Inadequate or excessive illumination
3.2.9   Inadequate Personal Protective Equipment
3.2.10 Inadequate ventilation
3.2.11 Poor housekeeping
3.2.12 Pressurized Containers
3.2.13 Slips and Falls

In reality, hazards are innumerable to be identified in one sitting. A hazard becomes a threat to the organization when an opportunity opens for it to approach or come in close proximity to the latter's resource. The development of threat in relation to the team's objectives becomes another challenge called risk.

SECURITY RISKS AND MISHAPS

Risk is created when despite the threat that a hazard poses to the team resources, the hazardous condition is not changed. But in spite of the seeming singular definition of risk, it connotes several other meanings which all belong in the realm of security.

Risk to insurers is commonly referred to as the cost of the resource exposed to hazard or danger against which the said subject resource is insured.

To finance people, it connotes a condition in which an entrepreneur pursues his investments with an objective of making it grow despite the uncertainty of the outcome.

To the risk management practitioner, it is the product of the impact or the value of the resources to be lost as a result of the hazardous event and the likelihood of a mishap to occur.

To one who practices safety, risk is estimated by the amount of time a particular resource is exposed to a specific hazardous condition that could result to incapacity of a human resource to perform his work.

The definition of risk that is closest to the application of security is that, it is the potential for damage, loss or harm to people, assets, environment or reputation. It is the combination of threat of a hazard, susceptibility, vulnerability and their potential impact on the asset.

There are two risk categories: **Financial Risks** and **Operational Risks.** Their basic difference is that financial risks considers basically quantifiable losses in terms of money due to obtaining conditions in the environment, economy and market place; while operational risks include those which are caused by adverse internal conditions that affect operations and it may also include demoralization, values and reputations, in addition to quantifiable losses. Another difference is in the process of determining their likelihood of outcome. Financial risks considers only factual and historical data; while operational risks are determined by many other factors such as obtaining condition, environment, geography, sophistication, state of morale, values, ignorance and apathy among others. For purposes of understanding all threats, hazards and risks that threaten the resources of the organization shall be collectively referred to as Security Risks.

The event or outcome after a threat is realized when the hazard has interacted with an asset or resources resulting to a loss, damage, injury or death are called **Mishaps**.

THE CONCEPT OF THREAT DIFFERENTIATION

Threat Differentiation is the process of identifying and isolating the different elements that constitute a threat or a mishap and its sub-elements. The sub-elements should be treated and neutralized first before the preceding element can be neutralized.

All threats and mishaps have three elements. These are the **hazard,** the **target resource** and the **opportunity** for the first two to meet. Mishaps occur when these three elements are all present in a given situation. For example, the elements of a robbery are the robber, the potential victim and the opportunity. If any of these elements were not present there can be no robbery at all.

The robber as an element has three sub-elements, namely a **person**, his **capability** and his **motive**. If any of these sub-elements were removed there could be no robber and therefore there could be no robbery.

Opportunity is the instance under a certain condition where the robber and the potential victim will meet. Opportunity also has three sub-elements. They are the time, place and the obtaining conditions.

As in the case of robbery, if the robber and the potential victim did not meet at a certain time or place, there could be no robbery. However, even if they met at a certain place and time, if the obtaining condition were such that the place of meeting were well-lit and crowded, or that there is a policeman nearby, there could be no opportunity for the robber to perpetrate his act as the obtaining condition deters the perpetrator

The potential **victim or target** also has three sub-elements. They are the **person, susceptibility** and **vulnerability**. Susceptibility and vulnerability are often deemed as synonymous. In the practice of security, they are different from each other.

Susceptibility is the predisposition of an object or person to a threat. It often reinforces the motive or capability of the threat or hazard. For example, a political leader is susceptible to assassination while an ordinary worker is not. A business tycoon or his family members are susceptible to kidnapping while an ordinary employee is not. A resource may be susceptible to a mishap due to a specific hazard but it cannot be deemed vulnerable if it were protected with appropriate and adequate security measures.

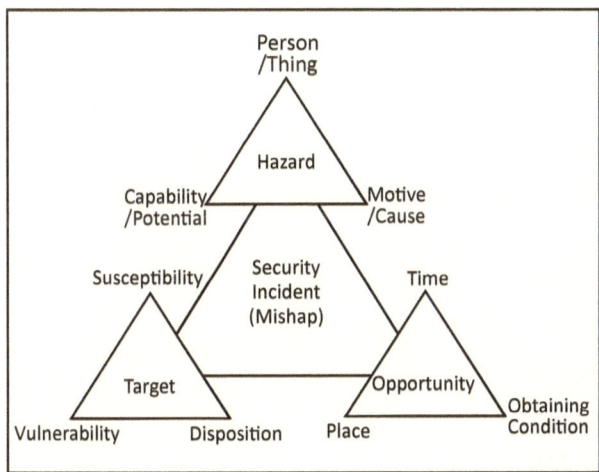

Figure 5. Security Incident Triangle.

A Security Incident has three elements: Hazard (person or thing), Target of threat and Opportunity. The process isolating and eliminating the elements of a security threat is called Threat Differentiation. A mishap will not happen if any of the elements were removed.

Vulnerability is the condition of a resource being exposed to a specific security risk or hazard. It is determined and measured by the presence or absence of a security measure to protect it. Company resources have varying degrees of vulnerability against a single security risk. In the same way, a single resource has varying degrees of vulnerability to different security risks.

The process of threat differentiation and the understanding of susceptibility and vulnerability are essential in the development and design of a security plan; it is also essential in minimizing the cost of security. Threat differentiation helps one to determine what specific element to address when neutralizing a threat. Knowing and understanding susceptibility will help one determine if a security measure or action is needed at all. This will be discussed further in Chapter 12.

THE BASIC SECURITY NEEDS OF AN ORGANIZATION

The basic security needs of a team should be defined in consideration of the team's resources and the security risks that threaten it. Thus, the following are the identifiable security requirements of any business, household and person:

1. Preservation and protection of resources against loss;
2. Maintenance of security measures;
3. Maintenance of a safe working environment;
4. Maintenance of facility and equipment conditions;
5. Accounting of all visitors and material movement;
6. Protection from own employees or members;
7. Accounting of all keys;
8. Compliance to policies and regulations;
9. Risk Management;
10. Crisis Management;
11. Upholding of its reputation

12. Redundant reporting of processes.

13. Feedback on condition of facility.

THE SECURITY FUNCTIONS CYCLE

The **Security Functions Cycle** is the never-ending security process of prevention, reaction and investigation. A security risk may remain to be harmless unless an ensuing action or condition has occurred. A security risk follows three stages of development for its effect to be realized. Each stage of development results to a certain condition. These stages are:

1. **Presence**. The first stage is the presence of security risk. The mere presence of security risk creates a threatened condition; in this stage there is no damage or loss incurred. We call this stage the "Threat Stage."

2. **Occurrence**. The second stage is the occurrence of consequent conditions or actions caused by a security risk or the encountering of a security risk. This stage creates an emergency situation. In this stage, damage or loss is being incurred. This event is called a mishap, which is also referred to as the "Emergency Stage."

3. **Consequence**. The third stage is the consequence or effect of a mishap. This stage is called the "Loss Stage."

At every stage of the development of security risk and the resulting condition or situation, a security action is necessary. These security actions constitute the Functional Cycle of Security. They are the following:

1. **Prevention.** This is the action to be done when a security risk threatens a resource. This is to prevent the occurrence of an incident caused by the security risk. This is the fundamental objective of security.

2. **Reaction**. This is the timely, necessary and correct action to take during the emergency stage when an untoward incident occurs. This is to avert the untoward incident or minimize the damage or loss of a resource.

3. **Investigation**. This is the ensuing action done after an occurrence of an incident caused by the security risk. This is to possibly recover the loss and extract lessons from the experience and formulate solutions to prevent the recurrence of the same incident.

The stages of the Security Functions Cycle indicate the need for an organization to adopt security programs to address the hazards that may put it at risk. These programs are widely accepted and indispensable concepts adopted by many organizations. The following programs, as a matter of principle, are functions of security:

1. **Risk Management.** This is a function of security that provides the guidelines to identify all the possible risk that may threaten the organization, assess the impact of these risks to the business and determine the likelihood of their occurring and their impact. The the impact or criticality of a risk is used to determine priority actions that the organization needs to undertake. It is by doing this process that the organization can determine the ways by which the risk can be addressed by elimination, avoidance or mitigation. The mode by which this function is performed is called constructive investigation or intelligence. The methods of Risk Management are similar to security assessment. This is discussed in detail in Chapter 12.

2. **Business Intelligence.** This is a body of knowledge about the environment, including the threats to the organization. It provides the organization information used as a basis for the planning and execution of its initiatives and goals. Out of business intelligence has emerged another concept

called **Competitive Intelligence**. The former covers all the elements of its environment such as natural, social, political, economic, physical and industry, while the latter focuses on the industry in which the organization operates and the different players in it as to their capabilities and limitations. Intelligence as applied to a specific threat is considered as the first line of defense of any organization.

3. **Crisis Management**. This is a function of security which provides for the policy and procedure to minimize the loss or destruction of the organization's resources resulting from an occurrence of a mishap. One function of Crisis Management is Business Continuity. This function provides for the procedure in ensuring the availability of essential resources to deliver the product or service to the clients in the event of disruptions.

4. **Case Management**. This is a function of security which provides for the proper disposition of all security incidents that cover not only the administration of sanctions to erring employees or apprehension of perpetrators of crimes, but also the determination of the causes of the disruptive events, drawing of lessons and the creations of means to prevent their recurrence. It is basically an extended form of reconstructive investigation. Reconstructive Investigation is discussed with some detail in Chapter 13.

The understanding of the Security Functions Cycle is fundamental to the application of a security measure or system. The value of the system is immensely enhanced if the three phases of the security function cycle can be executed. For example, it is generally accepted that a Close Circuit Television (CCTV) with a recording system is a combination of components that be used as security equipment. It is believed that an exposed CCTV camera can provide deterrence, say, for intrusion or theft. However, no amount of deterrence will prevent a determined perpetrator.

JOEL JESUS M. SUPAN

What the CCTV does is warn the monitor (assuming he picks up the warning while he is focused on other screens or attending to his other personal needs), to react to an event and record it. Any reaction to avert the commission of the intrusion or theft is always too late. Damage has been done. Of course, the recording may provide the team an image or identification of the perpetrator. But the end result is that damage has been done, which security should have prevented from happening. In this case, the CCTV does not address and satisfy the all the phases of the functions cycle of security.

The relationships of the above conditions and functions are further illustrated in the following table:

HAZARD	Presence of Hazard	Occurrence of Mishap	Consequence of Mishap
	Threat Stage	Emergency Stage	Loss Stage
Natural Hazards	=	=	=
Man-made Hazards	=	=	=
Sickness or Injury	=	=	=
Security Action	Prevention	Reaction	Recovery & Investigation
Management Policy	Risk Management	Crisis Management	Case Management

Figure 6. Security Functions Cycle Matrix. *Security Functions is a continuous cycle of activities namely, Prevention, Reaction and Investigation.*

THE BASIC PRINCIPLES OF SECURITY

To further understand security and how it is to be applied, it is basic for one to know its principles. These principles are the basic guides in the formulation of a security program. They must be considered and satisfied in the design of security systems. They were obtained and gathered from various security materials written by unknown authors in various security seminars and were validated by long years of experience in the practice of security. They must be understood by heart by every practitioner of security. The basic principles of security are as follows:

1. **There is no absolute security.** There are two conditions that exemplify this principle. First, it is that condition where hazards do not exist. There is no known place accessible to man or creation where hazards do not exist. This principle requires the need for the recognition that hazard in any form will certainly occur. The other condition that exemplifies this principle is that when a hazardous incident occurs, a resource or asset cannot be damaged or lost. A corollary to this principle is stated as, "Security is only applied to objects or resources with value." If an asset were subjected to absolute security it would lose its utility value due to inaccessibility or immobility. This is explained in the following formula:

   True Value = Relative Value X Accessibility

   True Value is that attribute of an asset by which it can be used for the purpose it is intended for. It is the product of an asset's relative value and its accessibility. True value of an asset can only be achieved if it is needed and it can be accessed, moved, used or perceived.

   Relative Value refers to the degree of value given to a resource or asset of the same kind and price by an individual. Everything is valued in a different way by different individuals.

JOEL JESUS M. SUPAN

To better understand this concept, take for example a ten-dollar bill. A ten-dollar bill to a business tycoon is loose change, but to a beggar, it may mean his food for a week, while to a trader, it may mean the price of his merchandise. Relative Value can also be attributed to an implement or tool in relation to its user. A hammer may be very valuable to a carpenter but it may not be as valuable to a musician.

Relative value can also be applied to two things of different worth in term of their price and purpose. To illustrate this concept, a telephone may cost less than a delivery van. They have different cost and different use but they may have equal value to a service company.

Accessibility is the attribute whereby an asset can be accessed or perceived to serve its purpose. Money has no value if it is kept in a vault that cannot be opened. A diamond, being primarily an ornament, has no value if it cannot be seen. A building loses its value if all its doors and windows were nailed shut to prevent intrusion and cannot be accessed.

It is this utility value of things that gives the reason for them to be protected or secured. It is the reason why things need to be secured.

This principle requires the need for the recognition that hazard in any form will certainly occur.

2. **Security is only as good as its weakest link.** Security risk is greatest or a mishap will most likely occur where there is least protection. Its impact can be greatest where there is no protection to the resources. Any weak point would be sufficient to neutralize or render ineffective the rest of the components of the security system. A steel chain is only as strong as a thread when one of its links is substituted with thread.

This principle requires that the strength of various security systems and measures used for a given asset requiring security must be evenly distributed around the asset being protected.

3. **High relative security can be achieved in depth.** Relative security is the average time for an average hazard to neutralize a certain protective barrier. Thus, a series of protective barriers would take a longer time to be neutralized than a single barrier. A high relative security can both provide protection and maintain the accessibility of an asset.

   This principle implies that redundant security measures are necessary. It also requires that each system is independent from each other. Each redundant measure must not rely on the other measure, but must be able to stand alone to produce the same degree of protection as the others. They should be supplementary rather than complimentary.

4. **Risks may come from within and without.** Security against external hazards does not guarantee protection. This is because most hazards come from within the team or its area of operation.

   To illustrate, no matter how effective the perimeter fence and the guard system a department store may have, if its own employees are dishonest themselves, such efficiency of the external security would be useless.

5. **No two installations are alike.** Every team has different circumstances from the other. Each has a different set of values and priorities. This difference is further extended to their environment and resources. This being so, they also have different security requirements. This principle requires that security systems and measure should be specific to the requirements of a facility so that different approaches to protection must be made.

6. **There is no impenetrable barrier.** Hazards are everywhere
    and they come in innumerable forms. No matter how deep
    a series of protective systems may be, it is just a matter of
    time that it can be penetrated or neutralized. This principle
    requires that security systems and measure must be constantly
    maintained, continually tested and upgraded as necessary. The
    Team Leader must realize that a security posture has to be
    adjusted according to the constantly changing environment
    and conditions.

7. **Security is everybody's concern.** Concern for security
    must be imbued in all members and employees, from the
    highest-ranking officer to the lowest-ranking employee.
    Employees refer to all workers regular or contractual who
    have inherent and routine access to the facility while serving
    the interest of the organization. It is upon this principle that
    a **Corporate Security Culture** is developed. This principle
    must also be imbued in all the members of the family. This
    principle requires that security must be explicitly stated as
    one of the duties and responsibilities of every member of
    the team.

8. **Deterrence is only good against the least determined
    perpetrator.** This principle means that deterrence is good
    only against the law abiding person but not to a determined
    offender. Deterrence has impact on the visual and the
    psychological attributes of a person. Thus, the close circuit
    television (CCTV) cannot protect a property form intrusion
    or loss. The next use of the CCTV with a recording feature
    is that it is a tool for investigation. It may be used to identify
    the perpetrator without a certainty that he will be caught. But
    with obvious certainty it will not prevent the loss.

9. **All aspects of security must be applied to secure an
    object.** The aspects of security shall be discussed in detail in

this same chapter. However, it cannot be avoided that they be cited to illustrate this principle.

Objects of security are resources or facilities that need to be secured or protected. The specific objectives and circumstances of these objects or facilities create a need for specialization or a peculiar security operation. But, regardless of these peculiarities of attributes and needs, all the aspects of security must be applied.

To illustrate this principle, examples of objects of security are the Chief Executive Officer of a company and the President of the country. The security required for their protection is a special security concept called VIP Security or Executive Protection. This is a specialized field of security application. But despite the peculiarity of Executive Protection, all the aspects of security must be applied.

This means all the official members of the office, the security escorts and all household personnel must be subject to the application of the Personnel Security Aspect and that all policies, procedure, rules and regulation that define the operation must be followed and enforced to ensure Operations Security. All Information regarding the VIP movement must be kept in confidence and must be disclosed only on a need basis applying the Information Security Aspect. Adequate physical security such as armed escorts and armored vehicles and fortified residences and offices must be utilized while the subject is in transit or even while stationary.

But, despite the similarity of the security measures to be applied to the two different executives, the degree or level of security measures applied are different. This is due to the difference in the significance of the offices where they belong, the hazards to which the executives are exposed to and the cost and practicality of such application.

Another illustration of this principle is Information Technology Security or what is called Information Security. In the subject of security these are two different concepts. Information Security is an aspect of security; while Information Technology Security is a subject to which all security aspects are applied. The object of security in the application of Information Technology Security is the hardware that is used to produce, store and retrieve information. The other object of security is the program that creates the information together with the hardware and the information itself that is generated, stored and recovered by the hardware.

Often you would see in the provisions of the security operation procedures on how to protect information, intricate password management protocols, upgrading policies, back up storage systems in and out of the facility.

What is generally inadequate in these policies are provisions on conducting due diligence for all the members involved in the operation of these equipment. Absent too in most of these IT Security Policies are the protocols for the classification of information generated as to their criticality and sensitivity to the organization. Moreover, you would hardly see any classification clearance system for the members of the IT unit. IT Security requires all the application of the different aspects of Security as well as the elements of Information Security.

Lastly, banks, buildings, factories, schools, hospitals, ships, sea ports, airlines, airports, trains, stations, mines, power generators, among others, are all objects of security and each have their peculiar security requirement but, common to all of them is the need for the application of all the aspects of security.

10. **The effectiveness of one aspect of security is dependent on the application of the other aspects.** Similarly with the preceding principle of security, the effectiveness of the Personnel Security Aspect (Policy) can be obtained only with the application of information security. All information about a team member must be classified and his records must be kept from open access by using steel cabinets with lock which are the applications of at least two elements of Physical Security.

THE ASPECTS OF SECURITY

To better appreciate the application of the foregoing principles and to appreciate the extent and significance of security in any organization, it is necessary to know the aspects of security. The aspects of security are basically the means and methods that constitute the security system. An aspect of security assumes the name of the resources it protects.

Further explanation of these aspects shall be covered in their respective chapters in this book. The Six Basic Aspects of Security are as follows:

1. **Personnel Security.** This is the aspect of security which ensures that the members of a team and those accepted to be members are capable, reliable, trustworthy, loyal and healthy. In a home, it means that every member of the household have the same values, every member of the family has an obligation to each other is healthy.

2. **Information Security.** This is the aspect of security which ensures that no classified, sensitive or critical information proprietary to the team are disclosed to unauthorized individuals, distorted, destroyed or lost.

3. **Operations Security.** This is the aspect of security which ensures that all critical policies, rules and regulation, systems and procedures are properly enforced and followed.

4. **Environment Security.** This aspect of security ensures that the environments where the organization exists and operates are supported. Environment refers to the natural, physical, social, political, economic and the industry surrounding the organization. It provides that the natural resources needed by the team are preserved and conserved. It means the compliance to all laws, rules and regulations imposed by the government that provides the team its legal personality and conduct. It is being aware and adaptable to the prevailing economic conditions. It means the orderliness of its physical environment and its good rapport with the communities and the citizenry it serves and the industry to which it belongs.

5. **Reputation Security.** This is an aspect of security which preserves the good reputation of the team, its brand and all the products and services it stands for. It ensures that the commitments made to the clients are served as promised. To a family and an individual, it is the family name and personal honor.

6. **Physical Security.** This a system of physical barriers placed between the hazard and the assets to be protected.

The Team Leader is responsible for the team's security. This is the aspect, which ensures that the Team leadership set the right direction of the team with due appreciation and utilization of the above aspects.

THE CONCEPT OF SECURITY COST EFFECTIVENESS

The common understanding of security cost is that it is a non-revenue generating expense outlay. This being the case, it is always the last to be considered in the budget and the first to be cut when necessary. Its value is hardly recognized as necessity when nothing happens. Yet is becomes the byword after a mishap where loss is incurred.

Such is security; the more effective it is, the harder it is to justify its cost. This mind set is reasonable, considering that the objective of the organization at the outset is to make money. Therefore, the less security cost is the more money goes directly to the bottom line. This mind-set brings about the dilemma of determining what is the ideal cost of security or the cost of ideal security.

There is no fast rule in determining the cost of security in absolute terms or in proportion to the revenue or operating cost of a company. This is because no two facilities are alike.

What is essential to know and consider is that security is an indispensable need of any organization and it is a necessary expense. The first step towards an effective security is to consider its basic principles. Those principles espouse that all threats must be covered and that there must be security in-depth. While its cost cannot be eliminated, it can be greatly reduced by the following approaches:

1. Develop an organizational security culture where every team member is an instrument of security. This may minimize the use of guards or surveillance cameras.

2. Integrate all security policies and procedures in to the operating procedures. This will provide focus on prevention of mishaps rather than reacting to them. The cost of reaction or recovery is very expensive and non-recoverable.

3. Take an active stance against the threat by knowing them before they can cause damage. This can be achieved with Security Risk Management. A gasoline station which is both susceptible and vulnerable to robbery replaced its expensive guard system and replaced it with the procedure of keeping minimum amount of cash in the register. They also installed a conspicuous sign which says, in effect, *"All we have in our cash register are loose change."* When the station was robbed after the adoption of the procedure, the cashier gave all the cash in the till that amounted to loose change. From then on, the gas

station has never been robbed again. Moreover, the owner was able to save on the cost of guards.

4. If physical security measure were essential and indispensable, bundle its cost with the cost of production. The amount should cover the cost of acquisition, maintenance and depreciation.

5. If the business were a service provider, explicitly charge the cost of security. Customers appreciate the visibility of security and they would be willing to pay for it.

CONCEPT OF CONVENIENCE AND COMFORT

The **concept of convenience and comfort** espouses the idea that effective security should not cause inconvenience and discomfort to its subjects.

One of the most prevalent obstacles in justifying security is the discomfort and inconvenience to which customers and operators are subjected to. It could slow down operation and touch into the sensitivities of the people as it sometimes intrude into the privacy or into the culture of people. These conditions often bear the sympathy of marketing people who would oppose the use of various security measures that would result to "**weak links**".

This condition can be addressed with communication, courtesy and creativity. It is important that sales people and more importantly, the customer know the rationale behind the use of every security measure. This can be communicated with posters, education of and spot briefing of all concerned together with a courteous operator or enforcer.

A good example is using sniffing dogs to detect explosives in a transport terminal in the absence of sophisticated explosive detection equipment. A thorough bag search is cumbersome, time consuming and intrusive. However, a sniffing dog can only be used for baggage and not on a person due to safety and hygiene issues. For the same reason, some people do not want their baggage to be directly sniffed by a dog. However, odor is communicable. The odor of explosives could be transferred to a person's hands and to the baggage. The same odor can be captured by swiping a tissue or a Teflon tape on the hands of the person or on the handles of the baggage for the sniffing dog to pick up.

An alternative procedure could be that a sealed container probed or punctured with a small tube attached to a vacuum cleaner. Once activated, the vacuum cleaner will suck the odor-contaminated air inside the container onto a filter, which will catch the odor for the sniffing dog to detect.

The development of technology for security operations has gone leaps and bounds in the past ten years. Human intervention was minimized since it's been proven inconsistent due to fatigue and boredom factor.

To answer this dilemma, x-rays to scan and detect threat items inside a baggage was invented. The use of X-ray machines eliminated the open bag search, which was very tedious, cumbersome, time consuming, intrusive of one's privacy and prone to missing out threat items. The old X-ray machines were designed to primarily detect firearms and other deadly weapons made of heavy metals and had distinctive shapes. Now, with the threat of terrorism where chemical explosives are used, the old types of X-ray machines became ineffective. Thus, a new technology in X-ray was developed where the images of different chemicals inside materials manifest different colors such as black for high-density metals, blue for light alloys, green for plastics and orange for organic substances. Since most explosive compounds contain substantial amount of carbon, their orange color make them easy to detect among other materials except fruits, food and clothing.

To further make it easy to differentiate the explosive from harmless materials, X-ray technology was further improved with the use of two-dimensional and three–dimensional x-rays to enhance images of the materials being scanned. Thereafter, the use of the Z-number was invented. In this system, the specific densities of materials and substances were predetermined and assigned with a constant called Z-number to differentiate explosive substances from common materials, whose images in the X-ray may have the same color. But this process requires considerable skill, which could diminish due to a lot of factors while being performed. Then, newer technologies such as the X-ray back scatter and the millimeter wave were developed. These new technologies lighten up and make explosive substances stand out from harmless substances.

As a result of this advancement in technology, scanning of baggage for threat items has become more convenient and comfortable for passengers and the X-ray machine Operators use of skill, fatigue and errors are minimized.

THE CONCEPT OF CONSTANCY AND CONSISTENCY

The **Concept of Constancy** implies the need for uninterrupted availability, the widest security cover and the reasonable number of redundancies to provide depth of measure against all types of threats. As what has often been said, "We need to be lucky all the time to be secured, but the terrorist need only one luck to succeed."

The **Concept of Consistency** on the other hand, implies the need for sustained vigilance in following security procedures. One good example is frisking or what is commonly known as body pat down. In an actual monitoring of performance consistency in a transport station where a breach test is routinely done, the friskers have the highest number of misses in detecting a threat item on the ankle. The reason for this is that the frisker has to bend down or squat to frisk the ankle. But after patting down over a hundred people, the physical fatigue of the leg and thigh sets in and the friskers could hardly bend any more, thus the inconsistency. This situation was subsequently corrected by simply asking the subject to raise the seam of his pants so that the frisker need not bend down low.

One of the prevalent challenges to this concept is "Dependency Syndrome." Dependency Syndrome is a condition that makes security personnel and operators of security machines let their vigilance down and become inconsistent as they depend on the security personnel before and after them. It also applies to people who let down their guard for the reason that they have installed supposedly sophisticated and expensive security equipment in their facilities. Security systems and processes therefore must be continually challenged to determine constraints to sustain constancy and consistency.

The risk brought about by dependency syndrome is compounded by the reality that the presence of real threats such as real explosive devises hardly ever happen. This condition can either lead to complacency or the absence of opportunity for learning to gain skill and competency.

PARADOXES OF SECURITY

The practice of security brings about concepts that apparently contradict the basic principles of running an organization. These paradoxes constitute the reasons why people find the rationale behind security difficult to understand let alone embrace as a culture. They are often the reasons behind some misconceptions on security. However some of these paradoxes are just apparent. A close look on these paradoxes would help one realize that they are not paradoxes at all because they do not really constitute conflicting concepts. They actually reinforce the need to understand and apply the basic principles and concepts of security. Some of the most common paradoxes of security are as follows:

1. For every advantage you get out of security, there is a corresponding disadvantage. An example of this paradox is a window grill that protects a home from intrusion would prevent the easy escape in case of fire. Another example is that a fence around a gasoline station can deter robbers but it also prevents easy access of customers who give it its revenues. Another example is Personnel Security which

requires that all applicants must be subjected to due diligence. But to have a through and reliable due diligence, an applicant must accomplish a long Personal History Disclosure Form and submit authentic document to support his disclosures. The effort to be exerted by the applicant to produce those documents is so much so that he will go elsewhere where such documents are not required. This paradox reinforces the principle of "weakest link", "security in depth" and "no two facilities are the same."

2. Security versus Safety. This paradox demonstrates the classic conflict of security with safety principles. The security practitioner says, lock that door to prevent intrusion; while the safety practitioner would say, do not lock that door to allow egress in an emergency. This paradox has given way to the design of fire exit doors with "panic bars," which just needs to be pushed to disengage the lock and yet it cannot be opened from the outside.

3. The product of security is "nothing". Other teams in the organization have a specific product or service to deliver. However, security is deemed to have done its job when it reports "nothing happened" at the end of the day.

4. The higher the cost of security the farther the organization is from achieving its goals. The ultimate objective is to make money, there is no point having more security if the organization is spending its resources for it. This paradox in some cases is manifested in the old conventions of security where excessive fortification is applied. Examples of these are the excessive use of guards with heavy weapons and the use excessive lights.

5. The more effective security is, the harder it is to justify its existence. Most organizations are inclined to believe that if nothing happened in the past, nothing would most likely

to happen in the future. Because, of this, they have the tendency to reduce their focus on security and cut down on their budget for security.

6. The importance of security is realized only after it has failed. This is manifested in all the happenings that we see every day. Some organizations are contented with the fact that nothing has happened and therefore their security is effective. But when confronted with the real threat security is often not prepared for to address it. The reality is that threats and the conditions change all the time. It is essential that security must be adjusted to address such conditions. They tend to be reactive. This paradox provides the rationale for the need for the concepts of consistency and constancy of security.

Summary

Back to the basics means knowing and applying by heart and by deed the fundamentals of security, which start with the mission of the team or a group or individual followed by appreciation of the assets or resources needed to achieve the stated mission and the hazards that threaten these resources. It means knowing and applying the principles of security with the use of the various aspects of security and knowing the factors that challenge those concepts.

-oOo-

SECTION II

The Aspects of Security

The Aspects of Security are the means, methods or systems which constitute the whole security system. No single security aspect can stand alone to achieve security.

This section provides for a detailed description of the basic aspects of security, what they are, why they are essential and how they are performed. The aspects of security as one would learn must be adopted and applied in all security applications.

CHAPTER 2

Personnel Security

WHAT IS PERSONNEL SECURITY?

Personnel Security is that aspect or means of security which ensures that the members whom the team accepts are capable, reliable, trustworthy, loyal to the team and healthy. It is also the means that keeps the team members healthy and protects them from danger or harm.

This chapter is especially for the Team Leader and the Human Resource Team. The Human Resource Team is responsible for building the membership of the larger team or organization. On the other hand, the team leader is responsible for the training, development and maintaining the morale of the team members.

Human Resource practitioners must learn by heart the entire security concept for them to appreciate, integrate and interface its functions with the other aspects of security as well as the functions of the other teams in the organization.

Personnel Security is as much a function of the Human Resource Team as the Team Leader. As such, they should have a clear understanding of the fundamentals of personnel security.

This chapter is intended to add to the conventional principles and practices of human resource management. It is also intended to give another perspective to the management of the human resource of the organization if not add a new dimension to it.

In considering personnel security, the following are the questions to be asked by the leader:

First, "Are the members' lives, limbs and properties in the workplace protected?" Second, "Are members of the team, capable, reliable, trustworthy, loyal and healthy?

It is upon these premises that the needs of the members must be protected from harm and the need for the team to be protected from its own members is defined. It is upon these premises that the need of the team for Personnel Security is conceived.

Personnel Security has two aspects. These are the "protection for the team members" and "the protection from team members."

WHAT NEEDS TO BE PROTECTED AND AGAINST WHAT

The members of the team are said to be its most important assets because they are the source of ideas, talents and skills to perform tasks.

The team members are the only resource of the company that has the faculty of intelligence, reasoning, communication, mobility and skill. Being made of flesh and having the above-mentioned faculties, they are the assets which are most exposed and most vulnerable to harm or danger. Any harm done to any team member will disrupt the operation of the organization or even cause irreparable damage due to loss of information or talent.

On the other hand, the same personnel with the same faculties could be the most destructive risk against the team's other resource. They can wittingly or unwittingly disclose classified information. They are exposed to temptation for personal gains. Their ineptitude or indifference can cause accidents or compromise product or service quality standards. Their weaknesses or vice can be the target of corruption or intimidation. Their environment in and out of the work area could influence their conduct and behavior.

Personnel Security protects the team from the loss of its properties from team member's ineptitude, ignorance, indifference, negligence, disaffection and misconduct.

Personnel Security is the best application for the security principle, *"Security should also be applied from within."*

Records show that more than three-fourths of all losses of an organization are due to team members and that the other quarter is by external causes.

These facts provide a very compelling reason for the organization to adopt personnel security as a priority.

THE ELEMENTS OF PERSONNEL SECURITY

The elements of personnel security are the means by which it can be achieved. The elements must be the basis of the Team's Personnel Security Program, spearheaded by the Human Resource Team. The implementation of a personnel security program is a component of the overall Security Program of the Team.

To appreciate better the elements of personnel security, let's review the basic functions of the human resource management. These are defining the job requirements of positions needed for certain organizational objectives, establishing the necessary competency requirements to perform the defined jobs, hiring suitable people to perform the said jobs and finally sustaining and improving competencies to address the inevitable and indispensable requirement changes of the job. Adjunct to this function is the management of the team member's morale and discipline.

The processes necessary for these functions are recruitment, hiring, orientation, training, supervision, evaluation, competency and character development. It is in these processes that define the elements of personnel security must be imbedded. The team Leader has as much responsibility as the human resource team in the execution of these elements.

The elements of Personnel Security are the tools by which the capability, reliability, trustworthiness and loyalty of prospective team members are determined and developed. The elements of Personnel Security are as follows:

1. **Personal History Statement.** This is a documented disclosure of one's personal circumstances, physical attributes, abilities and capabilities, experiences and personal references. This is also called a bio-data, resume' or curriculum vitae'.

2. **Due Diligence.** This is an activity involving the verification of all disclosed information and documentations.

3. **Education.** This element involves adequate indoctrination about the Team, its History, Vision, Mission, Organization, Culture, Policies, Regulations, Administrative Processes, Functions, Operating Processes, Services, Equipage, Facility and references.

4. **Training.** This element involves the complete orientation, assimilation to the organization and job requirements specific to the team member's position, as well as the organizational requirements and relationships for one to be effective.

5. **Personnel Management System.** This is the element which monitors the effectiveness of the team member in the delivery of his job requirement in support of the objective of his team and the organization as a whole. Monitoring could be through Performance Evaluations covering productivity, functions proficiency, readiness, efficiency and effectiveness. Personnel Development Program can be implemented to monitor and develop competencies.

6. **Leadership and Supervision.** It is the nature of man to need supervision. A supervisor is always the lynch pin for team member's productivity. It has been found in studies that one of the primary reasons why team members do not perform or leave their job is because of their supervisors.

7. **Debriefing.** This is a procedure where a team member shall be reminded of his commitment to keep to himself information

proprietary to the team.

One of the most crucial and critical tasks of the Human Resource team is the hiring of the most suitable person to occupy a particular position or to perform a particular job. Suitability does not only mean having the competencies obtained from education and experience, but also the character of the prospective team member.

APPLICATIONS OF THE PERSONNEL SECURITY ELEMENTS:

PERSONNEL HISTORY STATEMENT

Personal History Statement (PHS) is a written declaration of information about an individual applying to become a member of the team. This declaration is done by filling-out a standard form called the Personal History Statement Form. *(See Appendix 1. Sample Personal History Statement Form)*. This form is necessary to have a consistent and systematic means of soliciting personal information about the applicant. The use of this form is based on the following principles:

1. Know your people. A Personal History Statement is the first source of information about the prospective team members. It is in knowing ones' members that a leader can best utilize their ability and talent for productivity. This principle is a basic leadership principle being practiced by great leaders.

2. The more personal information a person declares about himself, the faster and the stronger his loyalty to the team will be. The Personnel History Statement is akin to "baring it all". It is where the first manifestation of candor becomes evident. Candor is a value that by itself fosters harmony and understanding between the members of the team.

3. The thoroughness of the solicited information may provide deterrence for applicants "with skeletons in their closet".

4. The Personal History Statement shall provide and facilitate the systematic, accurate and cost effective verification of the declared information.

5. It will provide the opportunity for the employees to write a systematic biography of themselves, which is not normally done by a person.

Components of a Personal History Statement (PHS)

The PHS must be comprehensive. The ideal PHS should include information from birth within the recall of the applicant. Next to ideal will be from the time the applicant started school. The declared information will used for subsequent investigation to determine the truthfulness and trustworthiness of the prospective team member. The Personal History Statement should have the following components:

1. **Personal Circumstance** (e.g. Name, Birthday, Place of birth) are information that must be counter checked using the applicant's authenticated Birth Certificate. Any deliberate inaccuracy may be an indicator of dishonesty.

2. **Physical Description or General Characteristics** (e.g. gender, age, height, built, race and complexion). The importance of the set of information would become more evident when a team member becomes missing.

3. **Next of kin within second degree of consanguinity.** This includes parents, spouse and children as those within the first degree of kinship and siblings who are within the second degree of kinship. This information may provide an insight of the team member's upbringing. The siblings are also good source of information about the prospective team member.

4. **Residence History.** This may provide insight to the character and reputation of the prospective team member. A

person who transfers residences within short intervals may be considered as someone having problems in paying rent or mortgage or that he may have difficulty relating with the neighbors. The present address must be verified by actual visitation. The neighbors can provide a lot of insight about the applicant.

5. **Education History.** This indicates the prospective team member's educational preparation. It also indicates the applicant's academic qualification, capabilities and competencies. The intermittent movement from one school to another also indicates the conduct of the applicant.

6. **Employment History.** This element is the declaration of all the employment and companies and the nature of his work and the duration of his tenure the in declared employer of the applicant. It includes projects and other works (personal and professional). This component is an indicator of the prospective team member's loyalty, his abilities, his autonomy and his relationship with superiors, peers and subordinates.

7. **Past Affiliations with Organizations** (Civic, Social, Professional or Church) and the degree of involvement, status and duration.

8. **Physical and Health Condition or Medical History**. These items of information indicate the physical condition of the team member at the time of entry into the organization. His physical well-being will have a direct influence on his productivity. This statement will also help the organization to preclude undue claims from subsequently disabled members.

9. **Skills.** This indicates self-regulation by the prospective team member as well as his competency and usefulness in other areas not necessarily on the job he is applying.

10. **Hobbies, Sports and Interests**. The regular engagements

in sports indicates a balanced lifestyle. That the candidate has no sports indicates the probability of the applicant to be stressed at work or the absence of modality in fighting stress.

11. **Travels.** When an applicant has a history of travels this indicate that he is a trusted individual in his team. Also, willingness to travel indicates his willingness to learn as it means having a broad perspective of one's environment.

12. **Financial Circumstance, Assets, Bank Accounts, Credit Cards and Liabilities.** This component, in relation to other components, such as job history and residences indicates how the prospective team member manages his finances. This is important to know as it may have adverse implications on his work and relationships with the other members.

13. **Personal affiliations and references such as friends, work, neighbor and objects of disaffection.** This is an application of the saying, "tell me who your friends (and enemies) are, and I will tell you who you are." Friends, associates and objects of disaffections are rich sources of information regarding the person's value and character.

14. **Law Enforcement Records.** Records of the police, the criminal investigation bureau and the courts would indicate any criminal activities. A candid disclosure would indicate willing reformation to become a useful member of the society. However, these records do not always indicate the true character of the applicant. A lot of people who have no criminal or citation records may have committed crime of moral turpitude and in the same light, not all those who have criminal records are inherently bad for anyone not to trust them.

15. **Handwriting specimen.** This is an effective way of developing a source or reference in any event that an issue of questioned document has been raised.

16. **Fingerprint and Handprint specimens.** This requirement has a strong deterrence against potential risks from the applicant. It provides for a good reference for investigation for identity verification.

17. **Sketch of Residence.** It is a generally accepted concept that you don't know the person until you have gone to his residence. Moreover, this requirement is best utilized when a team members is missing from work or has figured in an accident.

18. **Subscription and Signature.** The applicant's subscription and signature is a gesture that he is opening himself up to scrutiny and indicates sincerity and good character.

19. **Waiver for third party information verification.** The waiver is executed by the applicant to preclude any issue on privacy.

The PHS Form must have adequate and appropriate space to accommodate all the information being solicited.

DUE DILIGENCE

Due diligence is an element of Personnel Security which provides for the verification of the personal information disclosed by the applicant in the Personnel History Statement. It also provides for other sources of information that will enable the team to conduct a thorough appraisal of the applicant's qualifications, competency, suitability, reliability, character, reputation and other personal circumstances.

Due diligence is often times called Background Investigation. Good governance provides that a verification of the applicant's declarations must be done before he is even considered for employment. The extent as to how far back the investigation would go depends on the sensitivity of the position applied for.

Personnel Security provides that all job applicants must be subjected to due diligence. But to have a thorough and reliable due diligence, an applicant must accomplish a long Personal History Disclosure Form and submit authentic documents to support his disclosures. The effort to be exerted by the applicant to produce those documents is so much so that he will go elsewhere where such documents are not required. Another example is the use of a polygraph machine for pre-employment. While this is an acceptable and a standard procedure for some organizations, Human Resource staff would not subscribe to it and thus the subsequent high incidence of acts of dishonesty.

It follows that the higher the position is, the more due diligence should be exercised. Also, the applicant's experience and age would determine how far of applicant's history must be verified. However, this does not mean, that a person applying for a contractual messenger or housekeeper position will not be subject to due diligence. The principles of security dictate that all employees, regular or contractual, regardless of the nature of their jobs must be subjected to due diligence. The lowly paid janitors have access to sensitive and critical areas yet they are most susceptible and vulnerable to intimidation and corruption.

Basic Tools for Due Diligence or Background Investigation

The instruments are legal documents containing information that will be used as basis to determine the veracity of the information disclosed in the Personal History Statement (PHS). To establish the authenticity of these documents, the original must be required and presented.

1. **Personal History Disclosure Verification Form.** This is a tabulated checklist used as a guide for the systematic verification of disclosed information and reporting of the findings. This is simply a reformatted PHS that will put all items of information

for solicitation in one column. The disclosed information or declarations, which are the responses to the solicited information, are written in the next column. Findings or observations are in one column and the list of sources of the findings is in another column.

2. **National Agency Check.** To facilitate expeditious sourcing, the applicant may be required to submit certificates and clearances from various National Government Agencies. The information obtained from these agencies may not directly determine the inherent character of a person but any discrepancy between the disclosed information to that of the government issued certificate will spell an unsuitable applicant or may provide opportunity for the organization or the applicant to correct past inadvertence.

The above offices issue the following documents:

2.1 **Armed Forces or the Homeland Security Clearance.** This clearance certifies that the applicant is not a member of any terrorist, subversive or dissident group.

2.2 **Investigation Bureau Clearance.** This clearance certifies that a person has no pending case in any court of law within its functional jurisdiction. Once a person becomes suspect in a crime, the prosecutor's office will send a file of the case to the Investigation Bureau. The Clearance may have the following declarations:

2.2.1 **No Record on File.** This generally means that the subject person has no previous or pending case on record. It can also mean that his criminal record has not been filed.

2.2.2 **No Pending Case.** This may mean any of these two scenarios: that a person has had a criminal case but the court had it cleared, there was a satisfaction of judgment or the case has been dropped. If the person had previous cases these will be stated or enumerated in the clearance.

2.3 The Citizenship or Census Bureau Certificates. These certificates can be any of the following:

2.3.1 Birth Certificate. This may be a facsimile of the original birth certificate with a manifestation that it is a true copy. It will provide proof of the veracity of the applicant's declaration of his birthday, age and mother.

2.3.2 Marriage Certificate. This is a certification issued by the Census Office to attest that a person is legally married.

2.3.3 Death Certificate. This is a certification to attest that a person is deceased.

2.3.4 Diploma, Transcript of Record and Training Certification.

The above instruments are used to determine the veracity of the declared qualification of the applicant. The copy must be verified from their source. e.g. The school or the Government Agency for Education.

Building an Internal Capacity for Employee Background Investigation

Based on the principle of weakest link, every applicant must be subjected to due diligence. But one of the basic constraints that prevents the team from subjecting all applicants to due diligence is its cost. Often, the team would rationalize, " why would I spend for due diligence if I can't employ the applicant anyway?" Or sometimes, organizations believe that they will spend more for recruitment if more applicants were rejected due to derogatory or non-disclosure of critical information. Or, they would find it difficult if they required so much documentation from the applicants.

But what many organizations should realize is that hiring a potential thief is more expensive that investing on due diligence.

To minimize the cost of services for due diligence, a team can develop its internal capacity to conduct background investigation.

Here is what they can do.

1. Hire an investigator on a project basis or assign trained personnel to perform the information verification. The basic job is to determine the veracity of declarations on the Personal History Declaration and Verification Form. This can be done openly rather than clandestinely. The investigator or a security officer assigned to this task need not be licensed.

2. Have the team member or applicant accomplish the Personal History Statement and Verification Form in his/her own handwriting. Ensure that all questions are answered.

3. Have the applicant submit both original and photocopies of the following documents for comparison. A special request may also be done directly to the concerned government offices for the collective verification of your members or applicants.

    3.1 **Birth Certificate.** This certificate is used to determine the veracity of the declared name, age and birthday, mother and place of birth. This can be obtained from the Citizenship or Census and Statistics Offices.

    3.2 **Transcript of Scholastic Record.** This is used to determine the applicant's declared competency. This can be obtained from the Agency for Education or the Commission on Higher Education or its equivalent agency. For individual verification, this can be obtained from the school where the applicant graduated.

    3.3 **Police Clearance.** This document is used to determine the applicant's reputation and demeanor in his /her immediate locality. It should be obtained from the applicant's permanent residence and his/her latest place of work. Individual police clearance should be obtained.

JOEL JESUS M. SUPAN

3.4 **Investigation Bureau Clearance.** This document is used to determine if the applicant or team member has not been convicted or accused of crime in the court of law.

3.5 **Intelligence Clearance.** This document is used to determine if the applicant or team member were not or has not been involved or affiliated with any criminal syndicate, subversion, and secessionist or terrorist movement.

3.6 **Clearance from last employer.** This clearance is used to determine if the applicant has not been dismissed for cause. It is also used to determine the veracity of his declared competency and experience.

4. The investigator shall then compare the original documents with the declarations and put the verified information in the finding columns. The investigator shall indicate the source of the information.

5. The investigator shall place "confirmed" or "conflicting" under the appropriate   (remarks) column. All conflicting findings must be written in red and all confirmed findings must be written in blue or black.

6. For undocumented or inconsistent information, the investigator shall visit in person the appropriate reference person.

7. The investigator should personally go to the declared residence and confirm the sketch or vicinity map of the declared place of permanent residence.

8. The Investigator shall check on the following within the neighborhood:

8.1 Visit the nearest convenient store or look for an

undeclared reference (Throw-off), who may know the subject and ask for the reputation of the subject in the neighborhood.

8.2 Ask for known vices like drugs, drinking or gambling;

8.3 Ask for other people who are known to dislike the subject and interview these persons ad ask for their reasons for disliking the subject;

8.4 Visit the residence and interview the family members for further verification and advise them on the purpose of the neighborhood check. Look for any sign to determine his affiliations or sign of inconsistency in his declaration inside their homes;

8.5 Go to the Local Government Office to determine the veracity of the submitted clearance;

8.6 Submit the accomplished Personal History Declaration Verification Form. Place personal comments as deemed necessary.

Any adverse finding may or may not be used as a basis for qualification but they should be considered in comparison with other applicants. The benefit of conducting a good background investigation is that the organization will get to know the person they are hiring. It is prudent to discuss any adverse finding to the individual. Considerations given to the applicant could provide the seed for an employee's loyalty to the organization to grow.

EDUCATION AND TRAINING

As stated in this chapter, ignorance is one of the two most prevalent hazards in any organization as they cause accidents or costly errors, the other being negligence. This aspect of security requires that an employee should start in his work running. He must be equipped with the broad, basic and specific knowledge of the job for him to properly carry out his task.

This requires that all new team members must undergo a thorough orientation about the corporation, be provided with the tools and references on how to access a specific knowledge. He must be exposed to live situations. The orientation of the team member must include information about his job and team that he needs to know, why such items of information are necessary and how to perform his job. The following items of information are the basic and mandatory contents of an orientation course for a new member of the team:

1. **Complete official name of the organization.** The complete official name of any organization is a requirement of the brand. The significance of the official name of the organization is akin to significance of ones family name, country or citizenship.

2. **Complete official address of the organization.** The reason for this is the same as that of needing to know the complete official name.

3. **Nature of business of the organization.** By knowing the nature of the business, the team member can better appreciate the environment in which he will be living much his life

4. **Mission and vision of the organization.** Know the business mission and vision will make team member focus their initiatives, aligning themselves with the same mission of every member of the team.

5. **Organizational structure.** This information will provide the team member the information from whom he/she can get instructions or information or to whom he could report and pass on information. Information that requires expeditious actions often times need to be delivered with

timeliness to the unit or person concerned.

6. **Key officers of the organization.** The need to know this information is the same as knowing the organizational structure.

7. **Layout of the facility.** The efficiency and effectiveness of a team member can be fully achieved by him knowing the layout of the facility. This is so he would know where to go in case of emergencies or evacuations. He would also know where get the emergency equipment if he needed them. He would know where the restricted areas are and its restrictions, which are among the regulations he has to abide by and enforce. It is in knowing the facility layout that he could assist people who need help in getting directions.

8. **Policies and regulations of the organization.** Team members, whether contracted or organic, are subject to the rules and regulations of the facility they are deployed in. Likewise, one of the members' functions is to enforce these rules and regulations.

9. **Security and safety policies and procedures.** The team members are part of the security system of the organization. Security is required and implied in every aspect of the company's operation. Every employee has a part to perform in the security of the company.

10. **Personnel Administrative Processes and Forms** to accomplished.

11. **All emergency procedures.** One classification of security function is reactive. In most cases, the team members cannot be in one place all the time but they are expected to take the initial action in an emergency. As such, they must know the emergency procedures by heart; and they must know their specific role in case of any type of emergency to prevent further loss of life or property.

12. **Emergency telephone numbers.** The team members must have access to government security services in case of emergencies. As such, they must memorize or at least have in their possession or working area the contact numbers of the nearest police station, fire station, hospital and the key of the team or government security services officers. They must also have alternative numbers to call in case of the unavailability of the primary numbers.

13. **Locations and status of all fire stations.** This information is a requirement of the emergency procedure for fire.

14. **Unit orientation.** The team members must be familiar with the mission, functions, structures and how they relate to the other units to the organization.

15. **Basic function.** This is based on the Job Description and the organization's expectations from the new member.

The monitoring of the extent of the competency of the team members can be done by using tools like competency monitoring tools. Results of these tools must be translated to individual development plans of the team.

PERFORMANCE MANAGEMENT SYSTEM (PMS)

The Performance Management System is the measure used to evaluate the team member's capability and reliability. It is a gauge by which one can determine if the team members' performances are synchronized with the direction of their respective teams and that of the other teams in a larger organization. It determines the quality of the members' relationships with their superiors, peers, subordinates, customers and support organizations and vice versa. It also determines the members' productivity, efficiency and effectiveness on the job.

Example of this system is the Performance Evaluation covering relationships, productivity, functions proficiency, readiness, efficiency and effectiveness. A good complimentary system to Performance Management system is Personnel Shrinkage Analysis. This is a system of determining the changes in the character and behavior of an individual which cannot be captured by the PMS. The red flags for these changes are vices, uncomely habits, foul language and expressions, changes in lifestyle acquired after employment. Changes in family circumstances or the company the member keeps could also be among the factors for these changes. Employees must update their Personal History Statement every time there are relevant changes.

LEADERSHIP AND SUPERVISION

Man's orientation since birth is that he needs to be supervised. It is supervision that provides team members the guidance to the day-to-day performance of his tasks. The supervisor should provide the leadership thus the need to be well versed on the principles of leadership. It is only with these principles of leadership that the team members are properly guided. Often, team leaders lack certain competencies that are supposed to be the requisites for their promotion. A team leader should be well versed in theory and application of all universal competencies necessary for the job. The values, demeanor, disposition, conduct and behavior of the team members are greatly influenced by the values, demeanor, disposition, conduct and behavior of the team leader, regardless of their difference in character, orientation or calling.

The Leadership Principles

The following leadership principles are time-tested. These were aptly described in the publication of the U.S. Marine Corps on Troop Leading Procedures Manual. Great leaders, who endeared themselves to their men and gave them the inspiration to accomplish their missions and achieve their goals, have practiced them. They are as applicable in any organization, big or small, then and now.

1. **Know yourself and seek self-improvement.** As a leader, one must not only have the specific expertise and skills in the

performance of his job. He should likewise possess a broad knowledge of other fields related to his profession. He must also know how to manage his team under varying conditions. His proficiency should go beyond normal administrative work. He should be abreast with technology and he must have the skill to operate technical tools needed for his work.

2.   **Be technically and tactically proficient.** A leader constantly evaluates his self to recognize his strengths and weaknesses. He must understand his own capabilities and limitations. He strives to improve and master himself to include his strengths and weaknesses. By being proficient, a leader would know if his team members were doing their jobs correctly. Being proficient is the only way that he can teach his team members. A leader cannot teach what he does not know.

3.   **Set the example.** A leader shows his members that he can and he is willing to do the same things they do. He must be physically fit, well groomed and appropriately dressed. He should be optimistic. He conducts himself well so his personal habits are not open to criticism. He exercises initiative. By his performance he is the best member of the team.

4.   **Know your members and look out for their welfare.** A Leader puts the welfare of his members before his own. He should get to know his members. He must be approachable. He ensures that they have good living conditions and he assists them in getting the needed personal support. He should be concerned about their health. He encourages individual development.

5.   **Ensure that the task is understood, supervised and accomplished.** A leader must issue clear, concise and positive orders. He should determine if there is any doubt or misunderstanding of the mission. He ensures that resources are available to accomplish it. He does not over supervise nor under-supervise.

6. **Keep your members informed.** A Leader explains why tasks are being done and how he intends to do them. He ensures that the members are passing the necessary information. He detects rumors and stops them. He publicizes information about the team's successes. A Leader must keep the members informed of regulations affecting them.

7. **Train your members as a team.** A leader encourages unit participation in recreational and work related events. He ensures that the training is meaningful and has a purpose. Every member should know the other members.

8. **Develop a sense of responsibility among subordinates.** A leader uses the chain of command in telling them what to do and hold them accountable. He gives his members the opportunity to perform at higher levels. He recognizes accomplishments. He accepts honest errors and supports his subordinates.

9. **Make sound and timely decisions.** A Leader develops a logical and orderly thought processes. He makes plans for every possible event as time permits. He considers advice and suggestions from the team members when possible. He encourages the members to plan.

10. **Employ your team in accordance to its capabilities.** A leader never volunteers his team for an impossible task. He assigns reasonable tasks to his members. He assigns tasks equally among the members. He uses the full capacity of His team before requesting assistance.

11. **Seek responsibility and take responsibility for your actions.** A leader learns the duties of his supervisor. He seeks different leadership positions. He takes opportunity that offers increased responsibilities. He performs his tasks, big or small to the best of his ability. He stands up for what

is right. He evaluates failure carefully before taking action. He takes the initiative in the absence of a senior and he performs as if they were present.

DEBRIEFING

Debriefing is a done before a team member leaves the team after his or her attrition. This is the role of the supervisor. It is basically a part of Information Security. It is a process, which ensures to prevent the undue disclosure of classified, critical and sensitive proprietary information, which the subject team member knows. It is during debriefing that non-disclosure of classified information and a non-compete statement is re-iterated.

The debriefing is done by making a pre-set question and a reminder of the policy of disclosure of classified information. The team member must be subscribed.

Summary

Some surveys show that eighty per cent of all losses in any organization are perpetrated by the team members. The other twenty per cent of losses happened because the team members did not do anything to prevent it. This is the single premise why it is essential for the team to know its team members and ensure that they know their jobs, they can be relied upon, trust worthy and loyal to the team. The elements of personnel security provide for the guidelines to ensure these qualities in all the members of the team. It is as much as concerned of the team leader as the Human Resource Team.

-oOo-

CHAPTER 3

Information Security

Information Security is that aspect or means of security that prevents the destruction, loss, distortion or disclosure of classified, critical and sensitive proprietary information to unauthorized individuals.

IMPORTANCE OF INFORMATION

Information is the life-blood of any organization. It is an idea created that leads toward achieving the organization's objective in the fastest and easiest manner with the least possible cost. It is conveyed from one team member to the other, electronically, orally and/or visually with the objective of accomplishing the organization's mission and achieving its goals.

A body of information that is written, stored and read physically or electronically is called a document. This is why in some applications information security is sometimes called Document Security.

Information can be likened to the human blood, which conveys the nutrients from the food ingested to the different organs in the body. If the blood is drained from the body, the latter dies in the same way that the organization dies if its information were lost. If the blood if the buddy is corrupted with toxic substances the body dies. If the information in the organization is distorted, Teams leaders will make the wrong decisions and the team will not achieve its goals.

Information can also be likened to the rivets, bolts and nuts that bind the parts of a car. The flow of information between team members binds the organization together. The result of information is the sharing of ideas or even feelings that further results to the

JOEL JESUS M. SUPAN

coordination of actions of the functional units of the organization. The conveyance of information is communication. Once communication stops between two units, the organization stops to function and it dies.

Information can be anything from the organizational structure and the identity of its key officers, the organization's plans, production methods, products, pricing, marketing plans, capabilities and facilities layout among others.

By knowing the key officers they can be susceptible to kidnapping. By knowing the company's intentions, the competitors can preempt the organization's initiative. By knowing the organization's capabilities, the competitors may plan to undermine it. By knowing the layout and security plans, the criminal mind may sabotage or steal from the organization. With the loss of information, the organization will suffer a disruption of the business processes.

Information is basically represented by a physical or digital document. A physical document represents information whose text is written in ink on a surface. Digital document represent information that is written, stored electronically and is read as light image on a digital screen.

It is on the digital document or information that IT (Information Technology) Security should apply the elements of information security are as follows:

THE ELEMENTS OF INFORMATION SECURITY

1. **Need to know.** This element of information security provides that only people who need to know are the only people who are allowed to have access to that information in any manner. The element is applied by providing clearances to individuals in the organization only on levels or classification of information they are allowed to have access to.

2. **Preparation.** Team members, regardless of rank and position or tenure can create classified information in the routine conduct of their functions. The basic requisite for the preparation of classified information is knowing "who needs to know" the information. Thus, the number of copies in the case of written information or of addressees in case of digital preparation must be limited to the number of persons who "needs to know." It is during the preparation of information that it is classified as to its criticality or sensitivity. It is also during preparation that the appropriate cover sheet or heading is attached.

3. **Classification.** Information or documents can be defined according to their sensitivity or criticality. Classified information is proprietary information whose disclosure, distortion, loss or destruction is detrimental to an individual, a unit or the entire organization.

   Sensitive information is the type where improper disclosure can affect the reputation of an individual, a unit or the organization as a whole whether or not actual loss of resource was incurred. Its impact is generally psychological or emotional that affects the morale of the members of the organization.

   Critical information, on the other hand, is information whose disclosure, distortion, destruction or loss has direct impact to the organization such as loss or damage of resources. Whether it gets to be known or not by many individuals it will impede the organization from achieving its goals.

   Classification of Information is a system adopted to classify information as to their criticality and sensitivity. The system of classifying information or document may vary from one organization to another. There is no hard and fast rule in classifying information except the system used by the military. Companies may adopt their own system. What is important is that it should be made as a policy and should be known by all team members. The following is an example of how information is classified.

3.1 **Unclassified**. This classification shall be used for all information for general circulation such as announcement of activities involving all employees, regulations or instructions, complimentary greetings and messages, etc. Their disclosure will not have or have negligible adverse impact to the organization, its members or to the public.

3.2. **Confidential**. This classification shall be used for all information whose disclosure shall have moderate adverse impact to an individual or a small unit in the organization and their respective interests. Their disclosure shall be limited to those who have individual, functional or organizational (within the chain of command) concern. Specific authority from the unit head or owner is required for any one outside the chain of command to have access to the said information for every instance of access.

Access to these information or possession of its records by anyone outside the chain of command shall be documented in an access logbook specifically designed for that purpose. Examples of these items of information are the individual employee 201 file, reports of employee misconduct and its disposition or adjudication, routine operations journals and logbooks, physical development plans and other like information.

3.3 **Company Confidential**. This classification shall be used for all information whose disclosure shall have serious adverse impact to the organization and its interests. Their disclosure shall be limited to those who have individual, functional or organizational concern and the organization officers. Specific authority of the head of the organization is required for any one to have access to the said information subject to the prescribed clearance required by the policy. Those who have routine access to this information must be given prior clearance for this classification. Examples of these items of information are the corporate strategies and plans, financial statements, as built plans of physical assets, product or service development plans, marketing plans, product formula and like information.

3.4 **Private Confidential.** This classification shall be used for all information whose disclosure shall have critical adverse impact to the organization and is interests. Their disclosure shall be limited to those identified to have direct concern in any level.

Specific authority of the organization head is required for any one to have access to the said information. Those who can have access to this information must be given prior clearance for this classification. Examples of these information are investigation reports of incidents inimical to the company's interests, revolutionary concept papers for service or product development, extraordinary information that will put the organization to the advantage or otherwise.

As in other classified information, access to all Private Confidential information or possession of its records by anyone outside the chain of command shall be documented in a classified information access logbook specifically designed for that purpose

4. **Marking.** All classified information must be marked with their corresponding classification. For printed classified information, the classification markings are placed in bold red letters on the cover sheet and at the top and bottom of all the pages of the document. Different colors of the cover sheet are used for the classifications. Digital Programs for office applications have specific functions or features in applying and marking the classification of the document. The marking can be placed before the document itself.

5. **Transmittal.** The conveyance or flow of information is communication. Once communication stops between two units, the organization stops to function and it dies. Thus it is indispensable

that classified information no matter how critical or sensitive would have to be transmitted. The following are among the best practices for the transmittal of information. The following convention may be adopted in the transmittal of document hard copies:

5.1 The document must have a cover page with its designated color for a specific classification and the classification written in bold letters at the top and bottom of the cover. The cover shall not bear the title of the document but it shall bear the regulation on the handling of such document. The cover shall also bear the designated number of the copy being transmitted.

5.2 The document must be placed in a sealed envelop with the marking of the designated classification.

5.3 The sealed envelope is then placed in a sealed waterproof receptacle bearing the name of the addressee and the designated classification.

5.4 Only a person with clearance to handle classified document shall be allowed to carry the document to the addressee.

5.5 A transmittal form shall accompany the document to be signed by the addressee. The accomplished transmittal form shall be returned to the sender. Prior the age of digital technology, classified information is done in hard legible copies. Their transmittal requires that they have to be covered with a cover page specifically colored and marked with the appropriate classification. There is no fast rule on the format of the cover page so the team can design and adopt their format.

6. **Destruction.** Classified information must be disposed of. It must however be covered by a policy procedure that should cover retention periods, method of destruction and recording.

INFORMATION SECURITY AND INFORMATION TECHNOLOGY SECURITY

Information Security and Information Technology Security are often misunderstood to be one and the same–they are not. Information Security is an aspect of security while information Technology Security is a security system applied specifically for digital information. As such it requires the application of all the aspects of security.

-oOo-

CHAPTER 4

Operations Security

WHAT IS OPERATIONS SECURITY?

Operations Security is the aspect of security which requires and ensures the compliance, use and enforcement of all organizational policies, systems, procedures, rules and regulations. Organizational Operations are all the activities and undertakings of the different functional units of an organization with the end in view of achieving the organization's objectives. Each unit or element of an organization has a specific role and function to perform that contributes to the attainment of the objectives of the organization. Operations security is therefore the function of the team leader.

ELEMENTS OF OPERATIONS SECURITY

A leader must have a clear understanding of the following terms that constitute the elements of Operations Security. The elements of Operations Security are as follows:

1. **Policies**. Organizational Policies are the expression of the intent of the leaders that would provide guidance to the actions of the teams and individual members to achieve the set objectives of the organization. They generally cover the rules on how resources are to be utilized optimally according to their purpose.

2. **Procedures**. A procedure is a series or a sequence of set actions or activities that are performed in a specified manner so that the same expected result is obtained given the same surrounding circumstances.

3. **Systems**. System is a concept which aligns and relates policies and procedures into an operating process towards the smooth and seamless execution of the said procedure to achieve a desired result with the least use of resources.

4. **Rules and Regulations**. Rules are sets of established mode of behavior for the team members to follow. They are designed for the team members to keep a straight line towards the Team's objectives. Regulations on the other hand are used to keep the actions of employees in moderation.

Operations security can be best illustrated by a vehicle going on a trip. A vehicle's movement toward its destination constitutes a trip. A vehicle is used to reach one's destination at the shortest time possible.

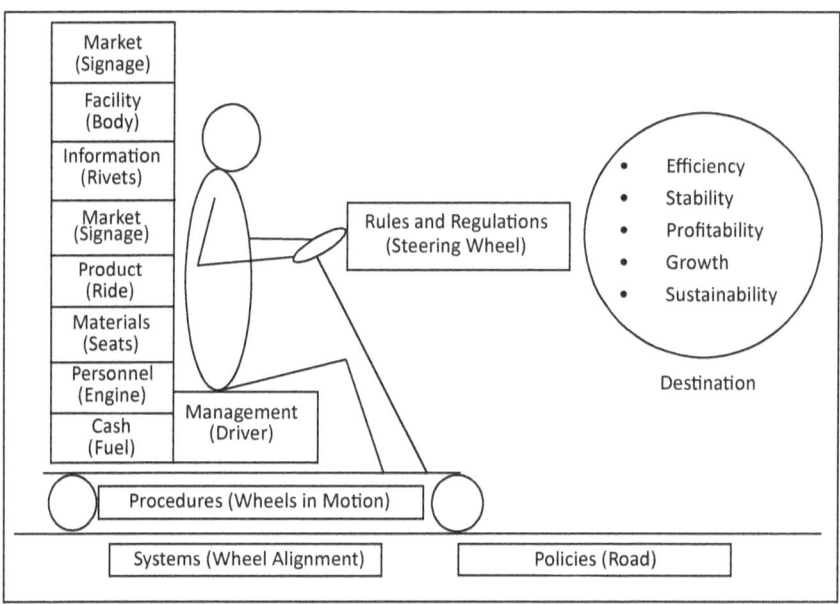

Figure 7. Operations Security Illustration. *A Business Organization can be likened to a person going on a trip. Every aspect of the organization corresponds to an engine part with similar function, all of which are needed to arrive at its destination.*

The vehicle represents the organization as a whole. The driver, representing the management, stirs the vehicle to its destination, which represents the team's goals. The vehicle parts constitute the organization's resources. Thus, the body of the vehicle represents the physical facility of the organization. The engine represents the organization's employees, who make the organization run. The engine runs with gasoline, which represents the various motivations of the employees such as their salaries, incentives or privileges, which are in form of money. The signage represents the market and the seats are represented by the services and the passengers are the customers.

The road represents the policies that guide the organization to its destination. The wheels represent the procedure performed by the employees. The alignment of the wheels represents the system, without which the wheels would not function well to bring the vehicle safely to its destination. The steering wheel represents the rules and regulation, which keep the wheels on the road or bring them back on the road, should they wander off.

CHALLENGES IN COMPLIANCE AND THE REQUISITES OF OPERATIONS SECURITY

Compliance to Policies, Systems, Procedures, Rules and Regulation is one of the challenges faced by the leaders in running the organization. These challenges are brought about by either the member's ignorance, indifference or both. Both are also hazards that threaten the organization. The threat of these two hazards range from the insignificant to the ultra risks. There are three requisites needed to attain Operations Security. These are:

1. **Education**. Two of the basic principles of Leadership are "Clarity of the Mission" and "Keep the members informed." These require leaders to ensure that all members of the organization know and have a clear understanding of all organizational policies, rules and regulations. They

must know why said policies are necessary and how they are enforced.

All team members must also know all the systems and procedures that are applicable to the entire organization and those which apply only to their respective areas of concern.

To enable team members to acquire the knowledge and understanding of the policies, systems, procedures rules and regulation, it is important that these are well documented and cascaded. Documentation is necessary so that the members will have access to information when they need it. Note also that Education is one of the elements of Personnel Security.

Since it cannot be expected that all team members will memorize all information the company requires them to learn, a program for cascade and education and a system of dissemination and publication must be adopted. The program must be implemented for all new hires and a periodic review and improvement must also be done for all existing members. Continued education can be reinforced by systematic publication by any medium.

2. **Engineering**. The other requisite for Operations Security is engineering. Engineering in the context of Operations Security means the provision for physical support or implement such as infrastructure, tools, among others to enforce Policies and Procedures.

3. **Enforcement**. It is in man's nature that he needs to be reminded of things he must do. Enforcement is a function of the leader. This is why it is important that the he must know all policies and regulations that he has to enforce. Effective enforcement can be achieved by making enforcement one of their duties, to look after one another and by providing them appropriate channels through which feedbacks can be transmitted to the management.

TOOLS FOR OPERATIONS SECURITY

In principle, the need for Operations Security was brought to fore following the adoption of the various best practices and standards in operations aspect of security by organizations all over the world. Among these best practices are as follows:

1. **Quality Standards.** The most widely known advocate of implementing standards is the International Organization for Standardization. ISO is derived from a Greek word "*isos*" which means equal.

    Standards ensure that all products and services have such attributes as quality, reliability, efficiency, environmental friendliness, safety, and interchangeability and produced at the lowest possible cost. These attributes are also the elements that support Reputation Security, which is one of the aspects of security.

    Another tool that has been adopted by some successful organizations to ensure the quality of the products and services as well as the internal processes in the organization is the Total Quality Management (TQM) designed in Motorola.

    Total Quality Management is a management approach that advocates and practices the concept of Total Quality. Total Quality on the other hand describes the attitude, discipline and culture of the organization in the production and delivery of their products and services to satisfy the needs of their internal and external customers.

    The discipline requires that there must be quality in all aspects of the organization's operations. When processes are being done right the first time, defects and waste are eradicated. Another popular tool for Operations Security is Six Sigma. This is a set of practice that was developed by Motorola to systematically improve processes to eliminate defects and ensure products are produced following set specifications.

2. **Policies and Procedures Manual**. The Policies and Procedures Manuals are tools to ensure that the dissemination of said policies is the same across the organization and will not be subjected to misinterpretations. Manuals provide reference and basis for future policies that would need to be changed on account of changes in the environment, conditions and requirements of the organization.

3. **Related References**. This component provides information about related policies or procedures, guidelines and forms of the company, laws of the country, specific laws, regulations or other references that are directly applicable to the procedure.

4. **Forms**. Forms are tools for enforcement. They provide simple and comprehensible instructions when accomplished. They should be easy to transmit, file and retrieve systematically. They can be appended tot he procedure. Some Security Guard Operations Forms are illustrated in Appendices 1 to 15 at the end of this section.

5. **Performance Evaluation System**. This tool is also one of the elements of Personnel Security. It monitors and ensures the individual capability and compliance of the team members to the set policies and procedures. It also determines their efficiency and effectiveness in the performance of their assigned functions and tasks.

6. **Operations Audit**. This is a tool used to determine the efficiency and effectiveness of a larger functional area of a team or a system. It also determines the full compliance to the set policies and procedures cognizant of the said team or system.

FORMULATION OF POLICIES, REGULATIONS AND PROCEDURES

One of the universal competencies that a team leader must possess is the ability to formulate policies and procedures. Organizations normally adopt a format in their preparation and presentation of Policies and Procedures.

Procedure should be simple, consistent, and easy to understand. In order to ensure a consistent format between documents, the organization must provide organizational templates to help leaders in writing policies and procedures.

The compliance or non-compliance to policies and procedures is often influenced on how they are written. Good Polices and Procedures must have the following attributes:

1. Policies and Procedure are written in clear, concise, simple and understandable language with the minimum words possible. They should be written in a manner that what needs to be done can be easily understood and followed by all users. Acronyms or technical jargons used are defined the first time they are used.

2. Policy statements provide the rule rather than how to enforce the rule.

3. Policies and procedures must be factual and accurate. Use of information such as name of designated people must be avoided, indicate positions instead.

4. Procedure statements are accessible and readily available to the team member.

5. Procedures should be developed with the customer and user in mind. Well developed and thought out procedures provide benefits to the user.

6. Involve users in the formulation and documentation of procedures to provide them with a sense of ownership.

JOEL JESUS M. SUPAN

COMPONENTS OF A POLICY PROCEDURE

Good policies and procedures contain the following components:

1. **Headers and Footers**. This component should contain the title, the date it was issued, the policy [procedure number, page number, date for the policy procedure to take effect, notification on which policy is superseded, issuing office, and the authority on policy procedure approval. The policy procedure number, page number and title must appear on all the pages. The footer should repeat the issuing date and the policy procedure title. These items of information are presented within a table at the top and bottom of the page.

2. **Policy Procedure Statement**. This is a straightforward statement of the general policy related to the implementation of the procedure.

3. **Objective, Purpose or Rationale of the Policy Procedure**. This component provide for the explicit statement of the objectives of the policy procedure. The objectives should take reference from the overall corporate objective. It provides for the context on which the policy procedure was made, stressing the rationale, importance and its significance to the objectives and interests of the organization. The policy procedure statement likewise, gives reference to other appropriate regulations relevant to the procedure on hand.

4. **Scope or Applicability**. This component identifies areas and activities and provides for the explicit statement to who the policy procedure would apply. It may also state the consequence for non-compliance.

5. **Definitions**. This component contains where certain terms that need definition to make the policy procedure comprehensible to those subject to it.

6. **Administrators, Concerned Offices and Help Desks**. This component describes and provides for the administrative requirement for assistance necessary to implement a procedure. This will also be used as reference for dissemination or cascading. The office and specific individual position title, telephone numbers and electronic mail address should be indicated in case they need to be contacted for interpretations, resolution of problems and special situations related to the procedure.

7. **Implementing Procedure.** This is the part of the policy procedure that states the step-by-step information on how certain activities should be done to achieve the objective of the procedures. There are digital program applications that are readily available in the market that can be used to assist one in the formulation and construction of procedures.

8. **Authority to the Policy Procedure**. This component should state the highest administrative or operations officer or group verifying the correctness and application of the procedure.

9. **Related Policy Procedure for Reference**. This component provides for information on related policies or procedures, guidelines and forms. List of complete references of the procedure and ensure that documents cited are readily available if needed.

10. **Rescission Clause**. This component states the provisions to rescind all prior policies that are not consistent with or in conflict with the new policy.

COMPONENTS OF PROCEDURE DOCUMENT

1. **Headers**. This component states the title of the Procedure, the date it was issued and the Procedure Number, Page Number, Effective Date, Notification on which procedure is

superseded, Issuing Office, and the Authority on the approval of the Procedure. The Procedure Number, Page Number should appear on all the footer of each page and should repeat the Issuing Date and the Procedure Title. The procedure title should be carefully selected so that it is simple and clearly conveys the procedure's content.

2. **Overview of Procedure Description**. This component describes the rationale and overall objectives, functions, or tasks that the procedure is designed to accomplish and the circumstances under which the procedure should be used.

3. **Areas of Responsibility and Ownership**. This component lists the units, offices, and individual job titles of positions responsible to establish daily control and coordination of the procedure, authority to approve exceptions to the procedure as necessary and procedural implementation, including responsibility for any required electronic or written forms. It sets the scope of responsibilities of departments, units, offices or individuals under the procedure, the procedural areas subject to discretionary modification (if any) and the responsibility for implementation.

4. **Detail of the Procedure**. This component identifies the persons subject to the procedure with the necessary procedural and "how to" information. It includes the definitions of unique terms to preclude different interpretations. This component also includes copies of all forms needed to document the compliance to the procedure. A transaction flow chart may also be included. Various approaches that are customized to the subject of the procedure can be used. It can be a statement in outline format of each step required, a checklist of what needs to be done, an explanation of how to complete the necessary forms or an appropriate combination of the above techniques.

5. **Related Reference.** This element provides information about related policies or procedures, guidelines and forms of the company, laws of the country, specific laws, regulations or other references that are directly applicable to the procedure.

6. **Forms.** Forms are tools for Operations Security in a way that they provide simple and comprehensible instructions when accomplished; and should be easy to transmit, file and retrieve systematically. Some Security Guard Operations Forms are illustrated with Appendices 1 to 15 at the end of this section.

7. **Performance Evaluation System**. This tool is also one of the elements of Personnel Security. It monitors and ensures the individual capability and compliance of the team members to the set policies and procedures. It also determines their efficiency and effectiveness in the performance of their assigned functions and tasks.

8. **Operations Audit.** This is a tool used to determine the efficiency and effectiveness of a larger functional area of a team or a system. It also determines the full compliance to the set policies and procedures cognizant of the said team or system.

-oOo-

CHAPTER 5

Physical Security

WHAT IS PHYSICAL SECURITY?

Physical Security is the aspect of security, which provides for a system of physical barriers placed between the risk and the asset to be protected. Physical Security can best be understood by the description of its six elements. It is however important that the principles and objectives of physical security be learned so that the elements can be appreciated.

OBJECTIVES OF PHYSICAL SECURITY

The objectives of physical security are to deter, detect, delay, deny, diffuse and document the threat, hazard or the mishap. These objectives also serve to indicate the capability and effectiveness of a physical security system.

These objectives should be attributed to every single physical security measure.

1. **Deterrence**. To deter as an objective of physical security means that the mere presence of a physical element serves to discourage the threat thereby preventing the untoward incident or mishap. This is objective generally applies to human threats. One example is a high fence. A fence defines the boundaries of the property. A high fence is a statement that a certain area is a private area and unauthorized persons are not allowed. It also implies that one will need to exert considerable effort to intrude a facility. Thus, a potential intruder is deterred by the high fence to intrude into the facility. Another example is lighting. A good source of

lighting will deter most potential intruder. An advertisement of the presence of other security systems will also serve as deterrence. Deterrence is a passive measure as it only has "to be" to have an impact to security as opposed to an active measure which can actually prevent the occurrence of a security incident. Good examples of barrier for deterrence are high fences, a well-lit facility or the presence of guards.

Deterrence, however, is often overstated as a security posture. In reality deterrence is effective only against the undetermined perpetrator. At best, it is only effective against peace loving and law abiding citizens.

2. **Detection**. A good physical security system must be able to detect a potential risk such as an intruder and should give out a warning to deter the intruder and warn the owners or authorities about the intrusion. A security system should also announce an approaching risk such as incoming natural hazards. Examples of barriers providing detection are energy barriers such as light for visual detection, Closed Circuit Televisions (CCTV), Passive Infrared (PIR) for detection of heat, vibration or magnetic contacts for disturbances and animals for the detection of sound, odors and images.

3. **Delay**. Perpetrators of man-made risks follow a definite timetable that they master before starting their operation. Any delay that the perpetrator encounters increase the chances of him being detected. This is where the principle of relative security is applied. Example of security measure that can be used for delays are fences, grills, locks and guard patrols.

4. **Denial**. This objective is achieved when a physical security measure actively and actually neutralizes the threat by overpowering it at the instance of the event. A good locking system or a guard is a good example of security tool that can deny an intruder. A CCTV combined with detection using a

remote microphone, a local loudspeaker and an alert guard is a good system to deny an intrusion.

5. **Diffusion**. This is the objective of Physical Security where the same threat from the same source has been neutralized with no possibility of recurrence. This is the ideal result of an effective security system. If applied correctly, it can also deflect the threat to another target and/or neutralize it. An example of physical security measure to achieve this objective is the presence of and positive action of guards. Another example is a combination of all security measures called security systems integration and convergence. These concepts of security are discussed in detail on Chapter 1.

6. **Documentation**. Documentation is the process of preserving the scene of an incident. It is an essential reference on the event in an ensuing investigation after a security incident had occurred. It is also a reliable source of information to determine security gaps and it helps find solution to prevent the recurrence of a particular security event. Examples of these physical security measures are Close Circuit Televisions (CCTVs) with audio-visual digital recording, tape recorders and cameras.

If we aligned these objectives with the three phases of the functional cycle of security deterrence and detection would correspond to the prevention stage. Delay would correspond to the transition point between prevention and reaction. Deny would correspond to the reaction stage and diffuse would correspond to the transition point between the reaction and investigation stages. Documentation would solely correspond to the investigation stage.

The following diagram illustrates the relationships between the objectives of security and phases of the functions cycle of security.

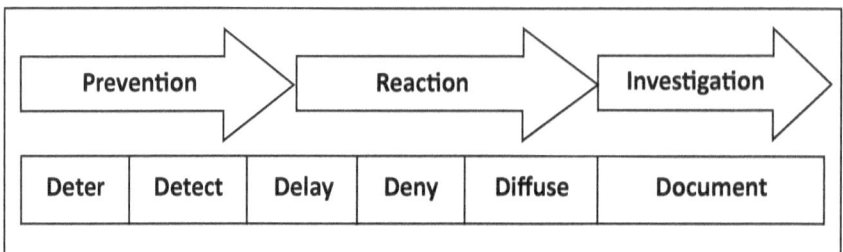

Figure 8. The Objectives of Physical Security. *The objectives of Physical Security each has a corresponding place in the Phases of the Security Functions Cycles*

REQUISITES OF A GOOD SECURITY SYSTEM

An ideal security system is that which can satisfy all the objectives of physical security. It could be the combination of components or a seamless interface of different security types, means or measures to provide the following:

1. Prevention by deterrence, detection and, diffusion

2. Reaction to delay, deny and diffuse a threat

3. Investigation to provide documentation for evidence as well as diffuse the threat by finding the cause and providing treatment

THE ELEMENTS OF PHYSICAL SECURITY

The elements of physical security are protective physical measures. They are applied to protect a facility from physical threats caused by nature as well as by man. There are six types of Physical Security.

1. **Animal Barriers.** Animal barriers are animals that are sensitive to changes in the environment. Their demeanor or disposition would send an early warning about an intruder or an impending disaster. An example is a sniffing dog, which is usually used for the detection of explosives and hazardous substances. Watchdogs and attack dogs are also good deterrence because they can serve as alarms. They can also

be used actively to deny or diffuse a threat. Animal barriers however, are not limited to dogs. Other domestic, farm animals or even wild insects can sense an impending danger.

2. **Energy Barriers**. Energy barriers include lights, sounds or electricity, which act as deterrence or detection of intruders. Security Hardware generally belongs to this element of physical security. Examples of these hardware items are Closed Circuit Televisions, Infrared Motion Detectors, and Magnetic locks.

3. **Human Barriers**. Humans are persons who are the well-meaning and friendly, within the surrounding of the installation. They could be neighbors, employees, guards, and members of law enforcement agencies. This barrier is the most prolific of all the elements physical barriers as they possess the human faculties, senses, reason and flexibility. It could be the most readily available substitute to most of the elements of physical security. They can deter, detect, delay, deny, diffuse and document threats and events.

4. **Mechanical Barriers**. Mechanical barriers are man-made barriers, but unlike structural barriers, they are movable or portable mechanical systems such as vehicles, locks and keys. They can basically deter, delay or deny threats.

5. **Natural Barriers**. Natural barriers are topographical features such as bodies of water, landmasses such as mountains, hills, wide plains and forests. These are basically passive defense as they deter and delay threats.

6. **Structural Barriers**. Structural barriers are man-made structures such as walls, fences, doors and window grills. They are basically used as substitute to natural barriers, which cannot be moved and are not on hand in all places. These are also passive defenses as they deter, delay, deny or diffuse threats.

LIMITATIONS OF PHYSICAL SECURITY

There are certain limitations that are to be considered in the application and design of Physical Security.

1. **Cost**. The cost of any security system should not be more than the worth than the potential loss that the organization can incur as a result of a mishap. To illustrate, suppose a warehouse has a capacity of USD100,000 worth of goods from which a profit of USD10, 000 per month can be earned, it is not practical to install a 24-hour guard system that would cost USD11, 000 per month.

2. **Function**. A physical security system can be obtrusive than can impede to the smooth flow of the operations of a team. Thus, it must maintain the accessibility or mobility of an asset being protected. For example, the reason why most department stores maintain open shelves is so that the merchandize are easily accessible to more clients. If they put the merchandize back behind locked shelves, they would need more sales attendants, spend more on structures and would have less sales.

3. **Location and Inherent Environmental Features**. A solid wall around a facility surrounded by wide open areas will provide cover to potential intrude. In the same manner a full view fence around a facility with limited and space and surrounding structures, would allow a potential perpetrator to survey the facility from the outside.

4. **Nature of Business**. It is not advisable for a gasoline station to put a fence or gate to protect it from robbers because it will also limit entry of its clients who want fast and convenient services

PARADOXES OF PHYSICAL SECURITY

The limitations of physical security are compounded by the paradoxes of physical security. Like the limitations of physical security, they must also be considered in designing security systems.

1. No single security measure can address all types of threats at the same time. For example a metal detector may detect a gun but it is not designed to detect explosives.

2. For every advantage there is a corresponding disadvantage. A physical security measure such as a window grill may prevent intruders from entering a house, but the same grill will prevent the homeowner from using the window as an emergency exit in the event of fire. Another example, the presence of light is deterrence to intruders but it may also make the target visible from a distance to a potential intruder. A high solid wall may deter a potential intruder in entering a yard but the same wall will prevent neighbors from hearing cries for help in the event of an emergency.

3. The cost of security should not be more that the value of the resource being protected or more than what one can reasonably afford to spend. One of the objectives of any enterprise is profitability and growth. An enterprise cannot grow if it were spending more for security than what it is earning.

4. Physical Security should maintain the selective accessibility, mobility and use of the resources being protected to maintain its effectiveness and value. A fence installed around a gasoline station will restrict its availability to motorist thus, limiting the business' growth or even its earning capacity.

5. Physical Security should consider environment and the use or function of the facility. A fence around a hotel would sacrifice aesthetics and landscape making it less attractive and unbecoming to guests.

SECURITY HARDWARE AND SOFTWARE

Security hardware items are instruments, tools, equipment, contraptions or other man-made physical objects or structures used in physical security.

A team leader must be familiar with these devices because these devices may become indispensable for the protection of the teams' resources. This familiarity, together with the knowledge of the principles of security and physical security, will make the leader arrive at an intelligent decision based on personal knowledge of the business rather than by the suggestion of the supplier. It may also result to acquiring a more efficient, manageable, effective and cost effective security system.

Security hardware is used to provide for the tools for the application of the types of physical security namely, animal, energy, human, mechanical, and structural barriers. A software program with various formats and applications drives high technology security hardware items.

Like the objectives of physical security, hardware is used to deter, detect, delay, deny, diffuse, and document intrusion in the facility.

Examples of security hardware items representing the types of physical security are as follows:

1. **Structural Barrier.** These are fences made from cyclone wires, barb wires, concertina wires or structural barriers like solid walls and vaults. These hardware items are primarily used for anti intrusion by providing deterrence and delay against a potential intruder. They also provide delineation of the facility. Other examples of structural barriers are body armor for personal protection and armored cars for the secured transport of portable and valuable item.

2. **Mechanical Barrier.** Locks and keys, combination and timed

locks for anti intrusion and access control can provide deterrence, delay and denial of threat. Other examples are vehicles, which are used to deter, detect and deny threats. Firearm is another example and it is used as deterrence and used to diffuse a threat. Turnstiles and revolving doors are used to control access in tandem with other access control systems.

3. **Energy Barrier**. Of all the elements of physical security, energy barrier offers the most number of types and examples. They are generally dependent on electricity and light as much as they are interdependent with one another to be functional and effective. Examples of this hardware are as follows

   3.1 **CCTV System**. This is a combination of a camera, recorder and a monitor. This system is basically used to deter, detect and document events. Earlier versions are analog CCTV, now there are digital based Internet Protocol (IP) Cameras and the recent network cameras. New technology now combines this system with other systems such as video analytic and facial recognition. It can also be remotely or locally monitored. CCTV can also be equipped with such features as a Pan, Tilt and Zoom (PTZ) so that a single camera can cover more area than its earlier version that is focused on a specific and limited area.

   3.2 **Electronic Detectors and Alarms**. These hardware systems are used for anti-intrusion by early detection and providing alarm to deter and diffuse the threat. Examples of detectors are Passive Infrared (heat) detector also known as PIR, motion detectors, vibration contact and magnetic contacts. These generally used for open or closed spaces. Where there are multiple detectors, a control box is interfaced with the system for a centralized control. The system can be easily set and reset or it can be programmed to activate and close for a specified time. The alarm can also be set in a remote area. This is usually in tandem with a central monitoring system. The following are examples of electronic detectors:

3.2.1   **Passive Infrared Detectors (PIR)**. This is a type of detector that would activate the alarm upon detecting heat coming from a certain mass of body such as a human or animal body.

3.2.2   **Magnetic Contact**. This is basically a switch that would activate an alarm or a camera if the two terminals attached to a door or window were either opened or closed. A normally close magnetic contact would switch on the alarm if the contact is broken, vice versa. A variation of this switch is a pressure contact where the terminals will come into contact to activate the alarm. This is normally used under the rug or carpet by the window or the door.

3.2.3   **Vibration Contract**. This type of detector is an actuator which is usually used for glasses, metal walls or wire fences. It activates the alarm when a glass is broken or the wall vibrates with a sudden movement. The sensitivity of this switch can also be adjusted to minimize false alarms.

3.2.4   **Light Beam (laser) for detection.** This is a switch that is composed of two terminals. It would activate the alarm if the light beam were disrupted. It is basically used for perimeter fences or walls. But, because of the intermittent false alarms due to events such as falling leaves, flying insects or small animals that disrupt the light, several beams parallel with each other at a certain distance are now being used. The simultaneous disruption of all of the beams is needed to activate the alarm.

3.3   **Alarm System**. This is the system that alerts people on the presence of an unusual event, an example of which is intrusion. There are basically four types of alarms. These are the sound alarms such as gongs and bells, the lighting alarms which are either fixed, flashing or rotating light, a public address system that is either one-way or interactive and a automatic telephone dialer.

3.4   **Fire Detection, Alarm and Suppression System**. This system is composed of a smoke and/or heat detector, an alarm system (bell), and an automatic suppression system. The sprinkler system is activated when the wax inside the sprinkler head melts due to the heat from a fire. The same heat will activate the thermostat switch and then activate the alarm. This system is generally incorporated in the building plan because it is required by the building code anywhere in the world.

3.5   **Air Analyzer**. This is a technology where the system is programmed to analyze the regular and normal composition, temperature and pressure of the ambient air inside a particular space. Any abnormal change could be detected by the system and will automatically trigger the alarm. This system is programmable to recognize regular and routine conditions and their changes to avoid false alarms.

3.6   **Lighting**. Lighting has long been used as deterrence for intruders. It is effective as a psychological deterrence.

Light can either be used as area light to illuminate a wide area or a spot light to focus on a specific space or area. The use of lights must also be studied and regulated because they consume a lot of power and the cost is high. As the saying on security lighting goes, "lack of light is risky while excessive light is waste."

3.7 **Access Control Systems**. These types of security hardware involve the combined use of identification, recording and automatic lock system. **Biometric Profiling** is a high technology system that uses the features or physical characteristics of humans to verify the identity of the person who wishes to have access into a facility. Biometric Profiling is based on the principle that no two persons have the same specific physical characteristic or feature. The systems all work by comparing the specimen, actually obtained on real time, from the person requesting access with the pre-identified and existing information in the database. All of these Access Control Systems require hardware for their application. Examples of these biometric profiling interments are as follows:

3.7.1 **Fingerprint Capture and Reader**. Fingerprints are peculiar to and attributable only to a single person. The system works by comparing the previously captured and recorded fingerprint or handprint and the specimen currently being read. The process is done by physical contact of the finger to a receptor.

3.7.2 **Finger Vein Identification.** This is a biometric system using the pattern and thickness of the blood vessel of a person's finger for identification.

3.7.3 **Hand Geometry**. This is a system of identifying the person by using the size, dimensions and the shape of his hand.

3.7.4 **Iris Scan, Capture and Reader**. This system of identification uses the features of the iris such as the color of the tissue around the pupil, the tissue pattern such as folds and spots. The pattern of the iris is obtained using a camera.

3.7.5 **Retinal Scan**. This is a system of identifying a person by using the patterns of blood vessels inside the retina. This pattern is unique to a person. This is obtained by using a low intensity light to read the patterns.

3.7.6 **Voice Identification**. This is a system of identifying a person through his voice. The system analyzes how the words are spoken combined with the parameters of the visual representation of the sound such as time, frequency, pitch, resonance and volume compared with the voice characteristics in the database.

3.7.7 **Pervasive Facial Recognition**. This system of biometric identification works by getting the features of a particular face and superimposing it onto a grid. The coordinates on the grid will lock onto the prominent features of the person and then captured and converted into algorithms. As result a subsequent template that is peculiar and unique to a person is produced and saved in a data bank. This process is called normalization. It is believed that the spatial proportion of the human face does not change after the age 8. This is the reason why the pictures of children could be recognized to belong to particular persons when they become adults. A specimen face that needs identification undergoes the same process of normalization before it is compared with the template in the data base. This type of technology is effective even when there is a distortion of the facial specimen due to a smile, a frown or other forms of facial distortion. The same efficiency is obtained if the face were tilted, turned to either side or up and down up to 30 degrees angle of facial displacement.

3.7.8 **Radio Frequency Identification (RFID)**. This identification system employs a microchip imbedded under the skin of a human or any moveable or portable physical asset. This is propriety and programmed to record any activity that is undertaken by the person imbedded with it. Its content can be scanned with a scanner.

The reliability of other types of access controls are often challenged, samples of these are cards with readable microchips and barcodes that are read by scanner to identify the person who bears them at close contact or proximity.

3.7.9 **Strike Latch and Lock**. This is a mechanical locking device that uses electricity to open or close and is activated by a switch. One disadvantage of this system is that in time, wear and tear will cause the barrel and the notch attached to the doorjamb to go out of line and will not engage to secure the door.

3.7.10 **Magnetic Locks**. These hardware systems are powerful locks using electromagnets with threshold strength of 300 to 600 pound per square inch and are activated by switches. One advantage of this over the mechanical strike lock is that it is not sensitive to door misalignments.

3.8   **Video Motion Detector and Video Analytic**. This hardware belongs to the more advanced technology systems used to detect and document events in closed or open spaces. It is also called video analytic that replicates a view of the vicinity in its memory and compares it with the same view at a certain time and it detects any change or discrepancy. There are a lot of variations of this hardware and just like camera; the development of technology is very fast.

3.9   **Weapons and Explosive Detectors**. These systems are used to deter, detect, deny, diffuse the introduction of weapons and explosives into a facility and provide for the documentation of its findings. These are commonly called scanners. Scanners are high technology machine used to visually scan target objects. Examples of these systems are the following:

3.9.1   **X-ray Scanners**. These are machines that are used to scan and provide x-ray images of a baggage's contents on real time. A metal weapon can easily be detected with the x-ray. The image of a metal weapon can easily be detected with the x-ray. However, the image of the more destructive explosive could be very difficult to detect when put together with other organic materials inside the same baggage. This is the reason why newer generation of x-ray machines are equipped with a system that exhibits different colors for different types of substances. Organic substances appear in orange color in the x-ray monitor. Heavy metal objects appear as black. Light metal alloys appear as blue; while plastics appear green.   But this system could not differentiate ordinary materials such as cloth, paper and food from explosives because they all have high percentage of organic carbon in them. Thus, the newer generation of X-rays are equipped with a Z-number counter system. This is based on the principle that each element has its own peculiar mass and density. The system designates a Z-number to identify each substance. A well-trained operator of an X-ray machine with a Z-counter system can determine explosive substances in the image shown in the monitor within a few seconds.

3.9.2   **Back scatter X-ray**. A new technology called back scatter is a variation of the x-ray. Where the conventional x-ray machine captures the transmitted x-ray through the target material, the back scatter technology captures the x-ray that reflects back from the target to form the x-ray image. The pattern of the back scatter is good for imaging organic materials, which will appear as a very light object in the image. It is also dependent or the properties of the material. One distinct advantage of this technology is that the machine is compact, mobile and it can cover larger and more objects for a shorter period.

3.9.3   **Millimeter Wave.** This technology uses extremely high frequency (millimeter waves) Radio Frequency. This frequency range is just below the (related) sub-millimeter "Terahertz Radiation" (or "T-ray") range. Clothing and other organic materials are translucent in some extremely high frequency (millimeter wave) radio frequency bands. This capability allows the operator to see any weapon inside the clothing of the subject. The millimeter wave is transmitted from two antennas simultaneously as they rotate around the body. The wave energy reflected back from the body or other objects on the body is used to construct a three-dimensional image, which is displayed on a remote monitor for analysis.

3.9.4 **Ion Mobilization System.** This system uses a technology that breaks down molecules to detect explosive substances. This technology uses the atomic or molecular composition on the principle of different explosive substances as well as prohibited drugs for identification. These unique characteristics are pre-stored in the system's memory. The machine converts a specimen into vapor by controlled heat, it then breaks down the ions of the substance and compare it with what is stored in the memory to determine the name of the substance. This system is said to be able to detect even with very minute amounts of substances.

3.9.5 **Chemo-illuminescence.** This system detects explosive substances that contain nitrates. The nitrate will chemically react with a colloid found in the system that will change its color to indicate the presence of the nitrate. Most explosives contain nitrates.

3.9.6 **Chemical Photo Analyzer.** This system is used to detect explosives by taking a picture of the substance. All substances have different crystalline characteristic and different ways of refracting light that results to a light pattern that is unique to a substance. These patterns are stored in the memory of the machines. For it to work, a specimen substance will be subjected to intense light. The light pattern that results will be captured and compared with the patterns, which are in the memory to identify the common name of the specimen.

3.9.7   **Metal Detectors**. This is device detects metals that passes within its range and sends off an audio and visual alarm to the operator. There are various models of this instrument from large and fixed frames known as walk-through metal detectors and hand-held metal detectors. This is used to detect firearms and deadly weapons. The presence of the metal detector itself has a deterrent effect.

3.10 **Investigative Tools**. Investigation is the last phase of the security function cycle before the cycle goes back to the prevention phase. As stated in the previous chapter, one of the primary functions of security is Investigation. Hardware items used for investigation are considered as security hardware. Examples of these hardware devices are the CCTV and the Polygraph (Lie Detector). The Polygraph is an instrument that measures specific physical but subtle reactions of a person such as blood pressure, skin moisture and contraction, pulse rate and intensity, the rate and intensity of breathing, change in pitch and timbre of voice and chemical change in the body fluids, when telling a lie. The abrupt changes in these vital signs are deemed to be signs of deception in the spoken statement of a person.

3.11 Other instruments for forensics used in investigation such as bullet comparators, microscopes, dyes, black lights among others, may also be considered as security hardware or gadget.

PHYSICAL SECURITY SYSTEMS INTEGRATION

Often those who need a security system are made to believe that a single security system can actually prevent the occurrence of the target risk.   But as previously explained the first principle of physical security states that no single security measure can address different threats at the same time.

One good example is a Close Circuit Television (CCTV) System. Users are made to believe that a CCTV is the solution to prevent unauthorized intrusion to a facility. However, in reality, all that the CCTV does on its own is to document the event; it cannot detect a threat. To be able to detect a threat it needs an operator to monitor the images captured by the camera. But even if the CCTV had an operator, it would have a very limited range of coverage and leave a lot of blind spots. The conventional solution is to add more cameras or install a PTZ camera. But then again, covering the entire facility can be expensive.

An operator is needed to monitor the images captured by the camera for detection. But the operators can only give attention to one image at a time. Thus, additional operators are necessary. But then again operators can only monitor on real time. Regardless of what they see, they cannot react on their own to actually prevent an incident.

The following tables illustrate how various security systems are integrated to satisfy all the objectives of physical security. They also show a series of security measures to provide redundancies. It is essential that these systems be supported by appropriate protocols and procedures to make them work efficiently and effectively.

JOEL JESUS M. SUPAN

1. CCTV Signage 2. Motion Detector 3. Audio Alarm 4. Central Monitor		1. Public Address System 2. Central Monitor 3. Guard Patrol		1. CCTV 2. Police	
Deter	Detect	Delay	Deny	Diffuse	Document

A	Threat Stage	Occurrence Stage	Consequence
B	Risk Management	Crisis Management	Case Management
C	Intelligence	Scene Preservation	Investigation

A	Condition	B	Required Action	C	Operating Mode

Figure 9. Integrated Physical Security System Against Intrusion. An Integrated Security System provides for cover of the Functional Cycle of Security with emphasis on prevention.

1. Crew Due Diligence 2. Anti-hijack Training 3. Deterrent Signage 4. Flashing Lights 5. Audio Alarms 6. PA System 7. Armored Shield		1. Armored Shield 2. Engine Shut Down 3. Cell Phone, Radio 4. Locks 5. Security Driver 6. Ramming Bull Bars 7. Vehicle Tracking 8. Reaction Team		1. Mobile CCTV 2. Police	
Deter	Detect	Delay	Deny	Diffuse	Document

A	Threat Stage	Occurrence Stage	Consequence
B	Risk Management	Crisis Management	Case Management
C	Intelligence	Scene Preservation	Investigation

A	Condition	B	Required Action	C	Operating Mode

Figure 10. Sample Integrated Physical Security System for Vehicle Against Hijacking An Integrated Security System provides for cover of the Functional Cycle of Security with emphasis on prevention. For vehicles susceptible to Hijacking, the primary consideration is the life of the driver and eliminating the fear factor.

SUMMARY

In reality there are hundreds of security hardware devices and systems that are used for various security applications. It is in knowing and better understanding how they can be appropriately used that would make them effective. Often the constraint for acquiring security hardware is its cost, availability or the frequency of use. One should remember, however, that no single security can stand alone to protect the facility from all types of risks. Security Systems composed of various pieces of hardware are generally available and it is through the resourcefulness of the team leader that these could be availed. Consider too the rapid development of technology for security. New technology could have been put to use by the time this book is published and circulated. It is therefore important that one has to continue to be on the lookout for new technology that can enhance the present technology to cover more types of threats at any given time, to address all the objectives of physical security, with lesser need for human faculties for their operation, less intrusive, less obtrusive and address new threats, all for a lower cost.

-oOo-

JOEL JESUS M. SUPAN

CHAPTER 6

Environment Security

Environment Security is that aspect of security where are all the elements that constitute the environment of the organization are addressed in a manner which will not disrupt the organization in the conduct of its business or in pursuit of its social responsibility. The following elements constitute the environment of any organization:

1. **Natural Environment**. This refers to gifts of nature, which supports all organizations. They need to be conserved, preserved and protected the environment to sustain the organization's existence. For example, a lumber company that does not conduct reforestation would eventually run out of wood to cut. To a broader extent, companies who do not treat effluence before releasing it to the sewer system would kill their surrounding environment. The policy makers in government have provided laws, rules and regulations to compel organizations to subscribe to these regulations. Non-compliance may cause huge penalties and worse, closure of the company. The sustained neglect and indifference towards the environment would bring about natural catastrophes that we are experiencing in these present times. These will eventually erode or destroy the company's resources. Standards for the care of the environment have long been established. While they are different from the laws on environment enacted and enforced by the government, compliance to generally adopted standards such as the ISO 1400/1401 (Environment Management), which was formulated by the International Organization for Standardization, has a positive impact on the reputation of the organization. Adoption and compliance to

said standard also facilitate the compliance with government regulation on environment.

2. **Physical Environment.** This element refers to the maintenance of the orderliness and cleanliness of the facilities of an organization. No client would ever want to patronize a restaurant whose surrounding is dirty. The upkeep of a facility's surrounding reflects the culture, collective attitude and professionalism of the organization and its members.

3. **Social Environment.** This element refers to neighbors, the members of the community and the society where the organization do their business and whom it serves. It is imperative for organizations to maintain its good relationship and rapport with its immediate community as well as the society. It is common, in these present times, that organizations have units that work to promote their Corporate Social Responsibility (CSR). An organization who does not give back to the society will eventually lose patronage

4. **Political Environment.** This element connotes, for lack of an exact word, the organization's need for a good relationship with the government that creates the laws, statutes, rules and regulations, and the duties and obligations that define the existence and regulate the conduct of an organization. By government, it means the national government, the branches and the bureaucracy that constitute it and the local governments who have their own statutes to enforce pertaining to the lawful conduct of business. The creation of the organization into a juridical person, payment of fees and taxes, compliance to the labor code, the building code, environmental regulation among others, are all manifestation of an organizations' healthy relationship with its political environment. The other element of political environment

is the compliance to laws, rules and regulation which the bureaucracy enforces.

5. **Economic Environment.** This element refers to the prevailing economic condition while the organization is operating. The basic law of economics that is the law of supply and demand affects all organizations. The ever changing availability of resources and the increasing demand can lead to the demise of any organization that cannot adapt to these changes . The organization must therefore be flexible in adapting to theses changes to be stable enough for it to achieve its goals. Protection and conservation of organizational resources is an essential undertaking to adapt to these changes.

6. **Industrial Environment**. This element refers to the particular industry and the market place for the organization in general. It is essential for the organization to maintain a good relationship with the members of the particular industry to which it belongs. This is notwithstanding its being a competitor in the market it serves. Industries often become a collegial body that assists the formulation of regulation pertaining to the said industry by the government. The sharing of individual and common experiences of the members of the industry can lead to the identification of the best practices. These practices will subsequently establish the standards that can be adopted by the individual members to improve the conduct of their respective businesses. A healthy industry environment could result to a strong and progressive industry.

On the other hand, the absence of a good relationship with the members of the industry would result to isolation that could impede the growth of the organization.

-oOo-

JOEL JESUS M. SUPAN

CHAPTER 7

Reputation Security

Reputation Security is that aspect of security, which ensures that the organization maintain a positive reputation to those that surrounds it.

Positive reputation is how the environment (elements of environment security) in which the organization operates, look up to it. The good reputation of the organization is built around its being able to deliver its commitment to all its stakeholders, customers, suppliers, the industry where it belongs, its employees, financiers and the enforcers of laws and regulations among many others.

Reputation is 'the result of what you say, what you do, and what other people say about you'.

Good reputation can be the most valuable and critical and sensitive asset of an organization.

Joachim Klewes and Robert Wreschniok who published the book Reputation Capital: Building and Maintaining Trust in the 21st Century wrote: *"reputation can be managed, accumulated and traded in for trust, legitimisation of a position of power and social recognition, a premium price for goods and services offered, a stronger willingness among shareholders to hold on to shares in times of crisis, or a stronger readiness to invest in the company's stock."*

The authors also said, *"Delivering functional and social expectations of the public on the one hand and manage to build a unique identity on the other hand creates trust and this trust builds the informal framework of a company."*

"This framework provides "return in cooperation" and produces Reputation Capital. A positive reputation will secure a company or organization long-term

competitive advantages. The higher the Reputation Capital, the less the costs for supervising and exercising control."

Not only that good reputation is expected from an organization but it is also expected from individual teams and its members.

While reputation is one of the six aspects of organizational security, its elements are composed of the same aspects for it to be achieved.

The good reputation of the organization can be achieved with the competence, commitment and integrity of its members, the integrity of its information, the delivery of quality service and products as a result of operations security and its good relationship and rapport with its environment.

Some organizations have established a position called the Chief Reputation Officer under whose oversight functions are the Brand Units and the Corporate Communications Units.

-oOo-

SECTION III

Guard System

Guarding System is one of the most common and conventional means of security. The guard, in fact, is so much an icon of security that it would be the first image that would come to mind when security is needed. The conventional mind-set makes guarding synonymous with security.

The person who performs guarding is called a Guard. Guards are, more often than not, referred to as "the security". By now, it should be well understood that a guard is just one of the three types of human barrier, human barrier is just one of the six elements of physical security and physical security is just one of the six aspects of security.

The team leader as the caretaker of the organization's resources should know the basic concepts and fundamentals of guarding because it is one of the means by which the resources of his team can be protected.

Based on the time-tested principles of leadership, "A leader must be technically and tactically proficient than his men." Thus, it is important that the team leader must also have a full understanding of the Guard System.

Even if the guard force were contracted from a third party, it is important that the leader must know what, why and how guarding is to be done. No one knows the security need of the team better than the leader.

Competency on the Guard System enables a leader to formulate the contract and the terms of reference including the level of service that he would expect from the guards and their provider.

This section will provide the learning on the basic functions of the guard and the different approaches on how to manage them.

CHAPTER 8

Basic Guard Functions

The guard is the most often used substitute for the other physical security elements. Most of the elements of physical security would require human intervention such as a guard to be effective.

If guarding were to be measured in relation to the entire security chain, it occupies only a very small portion of the said chain. However, despite its seemingly insignificant portion in the security chain, it is just as important as the other elements of the other aspects of security. This is because its absence would leave a void in that portion of the security chain and such void would create the weakest link upon which the integrity of the security system would be measured. While observation on the surrounding of the facility is one of the mandates of the members of the team, their focus on their respective jobs leave gaps, which the guards should cover.

Effective guarding involves time-tested principles. These principles are to be considered and appreciated by the user in the utilization of guards for security. Some of these principles are related directly or implied with the basic principles of security itself.

GUARDING PRINCIPLES

The basic principles of guarding are as follows:

1. A guard is effective only within the space that his five senses can cover. In fact, his human faculties limit his effectiveness in the same way that they are his basic tools. Because of these limitations, a single guard should not be expected to effectively prevent intrusion in a vast piece of land or even a much smaller area where there are a lot of blind spots that can impede his view.

JOEL JESUS M. SUPAN

2. The effectiveness of a team of guards is only as good as the most ineffective member. If several guards were used in a given facility to cover all the blind spots, the rest of the guard will be ineffective if one of them fell asleep. This is an application of the principle of weakest link.

3. The capability of a guard is limited to his human attributes. The capability of the guard to perform his functions is only up to the extent that he can carry a weight, run fast enough, have enough stamina to last and where his senses can be extended. Any normal person can function as a guard. Therefore, anyone can stand guard for as long as his human faculties are normal.

4. No two guards are alike. Guards, being humans, differ from one another. They differ in physical attribute, attitude and behavior when responding to stimuli, mind-set, disposition and a lot of other ways.

5. No two guard beats are alike. A beat is the specific patrol cycle in a given route in a specific period of time. A patrol cycle is one complete round of the guard's defined area of responsibility. The situation in a given area certainly changes with time so that appropriate adjustments have to be done by the guard for every patrol cycle.

6. Guarding alone cannot provide security since it is just one element of physical security.

7. The effectiveness of good guarding is measured by the non-occurrence of security incident. On the other hand, its efficiency is measured by the proper documentation of his performance of security functions and activities.

Anyone who needs and acquires the services of a guard must fully understand the above principles. He must also know the basic functions of the guards for effective supervision.

Guarding requires skills in performing specifically prescribed functions. It involves the use of human intelligence, physical senses, physical attributes, experience and flexibility to address a desired security need.

The functions of the guard are based on the universally accepted General Orders of a Guard, which was adopted from the General Orders of a Sentinel used by the military.

GENERAL ORDERS OF A GUARD

The general orders of a guard or sentinel provide for the principles upon which the Basic Functions are based.

1. To take charge of his post and all company properties in view;

2. To walk during his tour of duty in a military manner, keeping always on the alert, and observing everything that takes place within sight and hearing;

3. To report all violation of regulations and orders that he is instructed to enforce;

4. To repeat all calls from posts more distant to the guardhouse he is stationed;

5. To quit his post only when properly relieved;

6. To receive, obey, and pass on to the relieving guard all orders from company officers, officials, supervisors, post-in-charge or shift leaders;

7. To talk to no one except in line of duty;

8. To sound or call the alarm in case of fire and disorder;

9. To call the superior officer in any case not covered by instruction;

10. To salute all company officials, his superior in the agency; ranking public officials and commissioned officers of the military or police;

11. To be especially watchful at night and during the time of challenging, to challenge all person on or near his post and to allow no one to pass or loiter with out proper authority.

The foregoing General Orders, however, should not be construed as functions like most people and even the guards themselves do. The guards have to perform specific functions for them to carry out those orders.

THE BASIC FUNCTIONS OF THE GUARD

The basic functions of the guard provide for the manner by which the General Orders of the guards are applied.

The guard has the following basic function and activities:

1. Patrol the facility of the client
2. Record activities and observations
3. Report unusual incidents or conditions
4. Control and recording of visitors
5. Control access to restricted areas
6. Control and redundant recording of outgoing materials
7. Control and redundant recording of incoming materials
8. Control and recording of vehicle entering the facility
9. Control of pedestrian traffic
10. Control and oversight of parking spaces
11. Control, recording and safekeeping of keys and locks
12. Control of waste from materials, utilities and consumable resources
13. Control of hazardous materials and conditions
14. Gather of critical information to apprise management of the condition of the facility
15. Public relations
16. Assistance of the public
17. Enforcement of company regulations
18. Escorting of valuable persons and materials being transported

19. Manning the facility's communications network
20. React systematically and correctly to all emergency situations
21. Investigate security incidents
22. Help in maintaining peace and order in the facility
23. Proper turnover of duties at the end of his tour of duty

The basic functions of the guards are classified as preventive, reactive, and investigative. The contexts of these classifications are the same as the three phases of the functional cycle of security as illustrated in the Security model discussed in Chapter 2.

There is a lot more to knowing the functions. There are explicit objectives to the performance of such functions and there are methods by which the functions are performed. The safety maintenance and the wastage control are inherent to the patrol functions of the guard. The investigative function is covered in Section IV.

OBJECTIVES OF AND METHODS FOR PERFORMING THE FUNCTIONS OF THE GUARD

PATROLLING

Patrolling is an activity of a guard where he continuously observes the surroundings of the facility for any unusual situation or condition that can be inimical to its security and safety. Patrolling is the basic means by which the first General Order of the Guards is carried out.

For a guard to have an effective patrol, he must know why such patrol is being done, his general knowledge and specific circumstances of his area of responsibility and its components. Needless to say, a team must also know the same thing.

The Guard's Area of Responsibility and Its Components

The area to be patrolled is basically called the Area of Responsibility (AOR). The components of an area of responsibility are as follows:

1. **Critical Point (CP)**. Critical Point is a resource or valuable, whose loss or damage is detrimental to the organization. It is also a condition which can provide opportunity for mishaps to occur.

2. **Area Control Point (ACP)**. Area Control Point is a point in the AOR where the guard can have a maximum sensory coverage and reasonable physical control.

3. **Fixed Area Control Point (FACP)**. This is a point set on a particular location in the area of responsibility. It is upon which the guard shall record his location at a particular time.

4. **Transient Area Control Point (TACP)**. This is the actual location of the guard while on patrol.

5. **Control Area (CA)**. It is the area surrounding the Area control Point. Empirical demonstrations conducted show that a guard can be physically effective within a radius of fifty meters without obstruction. This area is also called area of liability because it is the area where a guard can control a subject within his capability.

6. Patrol Cycles is a complete round of patrol where all the established Critical Points were observed.

It is essential to guards to know the components of the areas of responsibility because they provide the basis for a good patrol plan as well as the extent of responsibility or liability of a guard.

Objectives of Patrolling

Patrolling has the following objectives:

1. To familiarize the guard of his area of responsibility (AOR) and establish its condition before and after his tour of duty and to establish his knowledge of its condition.

The first patrol cycle, which establishes the condition of the area of responsibility, will become the basis for the guard to determine if there were significant changes in the subsequent patrol cycle;

2. To show the guard's presence in his area of responsibility and provide deterrence to possible intruders or team members intending to commit violations. Intruders are attracted to abandoned or unattended facilities;

3. To ensure that the guard's area of responsibility is in order during his tour of duty based on what was established in the familiarization;

4. For the guard to verify and report unusual situation;

5. To gather information or evidence during an investigation.

How to Prepare for Patrol

Like all activities and undertakings, the key to the success and effectiveness of a patrol is in a good and proper preparation.

The first step in preparing for patrol is a thorough knowledge of the facility and its circumstances. The following must actually be learned upon the guard's assignment to the facility.

A team leader must ensure that a familiarization handbook is available to the guard for his reference. The guard on the other hand must seek out to know the needed information if it were not readily available to him. This information constitutes his stock knowledge of the facility.

Things that the Guard must know before Assuming Duty or Post

1. **Complete Official Name of the Organization**. The complete Official name of any organization is an integral part of its image. It is as important as the name of a person. Moreover this information can be put to use when screening

materials and correspondences being delivered to the facility. One of the red flags for suspicious correspondence of package is an incomplete, misspelled or wrong name of company. Such packages must be rejected outright or held in suspicion and must be subjected to further scrutiny of the sender. A policy for this must be formulated and enforced.

2. **Complete Official Address of the Facility**. The reason for this is the same as that of needing to know the complete Official name.

3. **Nature of Business of the Facility**. By knowing the nature of the business of the facility where a guard is deployed, he can better appreciate the threats and risks he must be aware of or the relative criticality of the resources he is required to protect.

4. **Organizational Structure**. This information will provide the guard the information from whom he gets instructions, or information or to whom he could report and pass on information. Information that require expeditious actions often times need to be delivered with timeliness to the unit or person concerned. Security is one function that may bypass conventional protocol or channels of communications

5. **Key Officers of the Organization**. The need to know these items of information is the same as that of knowing the organizational structure.

6. **Layout of the Facility**. A guard could never effectively patrol if he does not know the facility. He would not know where to go in case of emergencies or evacuations. He would not know where the sets of emergency equipment are. He would not know where the restricted areas are and the restrictions for which is among the regulations he has to enforce. He could likewise assist people who need help in getting directions.

7. **Policies and Regulations of the Organization**. Guards whether contracted or organic are subject to the rules and regulations of the facility they are deployed in. Likewise, one of their functions is to enforce theses rules and regulations.

8. **Security Policies and Procedures**. The guards are just a part of a larger security system. Security is required and implied in every aspect of the company's operation. Every employee has a part to perform in the security of the company as the guard has a specific security function to perform that compliments the others.

9. **All Emergency Procedures**. One classification of guard function is reactive. In most cases the guards being available in most places all the time in a facility, they are directed to spearhead the needed manpower to react to emergencies. A such he must know the emergency procedures by heart and as well as his specific role in case of any type of emergency, to prevent further loss of life or property of the company, as well of his own.

10. **Emergency Telephone Number**. The guard must have access to government security services in case of emergencies. As such he must have in his possession or must know by heart the contact numbers of the nearest police station, fire station, hospital and key company officers. He must also have alternative numbers to call in case of the unavailability of the primary numbers.

11. **Location and Status of all Fire Stations**. The information is a requirement to the knowledge on emergency procedure for fire. Their status must be checked every day.

12. **Location and Status of Telephone Lines**. This information is also a requirement to the knowledge of emergency procedures. In addition to the knowledge of the location, the guard must

know and establish the routine of checking telephone lines every day.

13. **Neighbors and their Security Conditions**. The security of any facility is influenced by the security conditions of the neighboring facilities. A neighbor that is vulnerable to crimes or fire also threatens the facility against criminals and fire. The neighbors also provide a buffer against any criminal activity.

14. **Security Records to Accomplish**. Security records as discussed in the succeeding chapter are documentations of the guard's performance of his assigned function. Records are the proofs of a guard's services.

15. **Critical Points of the Facility**. The guard must know all the assets or egresses that require attention. Not only must he know them by heart, he must also have a list to facilitate the seamless turnover of duty to another guard in case of routine and non-routine turnover.

16. **Patrol Area Control Points**. The guard must know the control points in his AOR. They will be the basis of the patrol plan which covers the route and the timing for a thorough and systematic patrol.

The next step in preparing for patrol is for the guard to ensure that he is physically fit to last his tour of duty. Patrolling is a physical activity and guard who is not physically fit may not last the patrol and may prevent him from performing his duties let alone his ability to react in case of emergency. Fitness is the part of the regimentation and discipline that a guard must maintain.

Lastly, the guard must ensure that he has complete, authorized and presentable sets of uniform including an identification card (ID). This is because the uniform is the symbol of the guard's authority and professionalism. It manifests reliability and respectability, which are essential in enforcing company regulations.

Tools and Equipment for Patrolling

The guard must ensure that he has in ready, appropriate and complete, organizational equipment. The following are the basic equipment for patrol:

1. Basic uniform
2. Nightstick
3. Whistle
4. Timepiece (must be synchronized with company clock and the other guards)
5. Writing pen
6. Notebook and patrol checklist
7. Flashlight
8. Service firearm (as required)
9. Portable radio (tested with control)
10. Camera
11. His five senses

How to Conduct a Thorough Patrol

1. The guard shall cover all the fixed area control points as listed in the Patrol Checklist.

2. He shall follow a pre-set pattern and a formatted patrol checklist.

3. The guard shall observe or look for the prevailing conditions (Critical Points) and check and/or correct any deficiency during the patrol.

What to Look for During Patrol

1. All unusually open, closed, or tampered gate doors and windows

2. All excess or lack of lighting

3. All safety and health hazards

4. All signs of wastage (water, power, materials, supplies, time)

5. All faucets and water closets for leaks

6. All electrical hazards (overloads, short circuits, loose connections)

7. All unattended heating elements, motors, appliances or equipment

8. Any misplaced object or objects alien to their surroundings

9. All trash cans and flower boxes

10. All unattended personal valuables

11. All classified information or materials

12. All visitors or personnel who are out of place

13. All emergency equipment and stations should be checked for completeness and readiness for use.

14. All scattered trash

15. All signs of weakness of building structures and leaks and flooding

16. All manifestations of vandalism

17. All violations of regulations

18. All signs of intrusion of man or animal

19. Any sign of infestation of stocks

20. All unusual sight, light, odor, temperature or sound

21. All the stocks within the route for anything unusual

22. Weather condition before and after patrol or any changes thereof

23. Any unusual change of sight, sound, odor, temperature in the premises or surroundings

How to Record Patrolling Activities

1. The guard must a have a Patrol Checklist using the prescribe format that would contain all the Critical Points (CP) and Area Control Points (ACP).

2.   The guard must record the time of start and end of his patrol in his journal.

3.   The guard must also record his arrival or departure from any control point.

4.   The guard must include the patrol results in the daily operations reports.

5.   All accomplished patrol checklists must be kept on file. A sample Patrol Checklist is attached in this Section as Appendix 2.

OBSERVATION AND DESCRIPTION

The guard is the extension of the eyes and ears of the team leader. He has access to most of the facility's spaces that are not routinely covered by the team members. More than just perceiving with his five senses, a guard must learn the techniques of proper observation for him to commit to memory, to easily describe and report his observations.

How to Observe and Describe Persons

The following are the basic techniques on how to observe and describe persons:

1.   Look for any outstanding or distinguishing mark, feature or mannerism.

2.   See if the person can be associated to a well-known person.

3.   Observe the general characteristics.

   3.1   Gender

   3.2   Age. Estimate the age then add and subtract five year from the estimated age to describe it.

   3.3   Height. Estimate height by using nearby references and then add and subtract one inch from estimated height to describe it.

JOEL JESUS M. SUPAN

3.4   Built. Built is simply stated as small medium large.

3.5   Complexion. State the color of the skin to describe complexion.

3.6   Race. State the race as Caucasian, Black, Malay, Oriental, etc.

4.   Observe his specific physical characteristics from head to toe like hair, forehead, eyebrow, cheek, chin, jaw, eyes, face, neck, shoulders, arms, hands, legs, feet, etc.

5.   Observe his variable features such as clothing and accessories.

How to Observe and Describe Objects

1.   Observe and describe its common name.

2.   Observe and describe its common use.

3.   Estimate its dimensions for regular shapes i.e. height, length, width, depth and thickness.

4.   Observe and describe relative size and shape of irregularly shape objects comparing it to the size and shape of known common objects.

5.   Observe and describe composition such as metal, wood, concrete, plastic, paper, etc.

6.   Observe and describe dominant color by comparing it with the color of common objects.

How to Observe and Describe Events

1.   Observe the effect of the event that made it unusual.

2.   Observe the place or location of the event, look for fixed landmarks.

3.   Observe the specific time of the event.

4. Observe all persons involved i.e. principals, victim, assailant, witnesses.

5. Observe and describe the sequences of the event.

6. Observe the factors or reasons that caused the event or mishap.

Practical Techniques for Patrolling

Aside from achieving the basic objectives of patrolling, a guard must also be on the lookout for his personal safety while conducting patrol. Thus, the following are suggested techniques in patrolling:

1. Do not establish time and route pattern that would make you predictable to the potential intruder.

2. Make a patrol route guide for each shift to ensure that all Control Areas, Control Points and Critical Areas are covered in every patrol cycle.

3. Backtrack as often as practicable to preclude the potential intruder from taking advantage of your past beat.

4. Keep both of your hands free as much as possible.

5. Do not pass steadily under the route lights. Doing so will make you visible and predictable to anyone observing you.

6. Do not walk too closely to walls and high fences to avoid being hit by falling objects.

7. When approaching a blind corner on one side, change your pace and sidestep away from the corners to prevent a surprise attack.

8. When approaching two blind corners on both sides slow down and watch for unfamiliar sounds, swerve to either side to cover the opposite corners and cross immediately to the other corner to cover it.

9. When entering a dark room, do not stay too long by the door. Otherwise, you will make a silhouette that would make you an easy target. Stay at the side of the door and make your eyes get used to the darkness before proceeding if you do not have a flashlight.

10. An intruder will instinctively make a source of light his target. Therefore, when using flashlight, keep it away from the body.

11. A lot more practical precautions can be learned in an actual beat. Just maintain a keen sense of observation and learn every lesson and share it with the other guards.

RECORDING OBSERVATIONS AND ACTIVITIES

Recording is the process of preserving information obtained by the guard while on duty and while performing his specific functions.

The two basic types of security records are reports and logbooks. **Reports** are used to transmit unusual observations. Examples of these are the unusual incident report and the investigation report.

Reports are also used to summarize routine operational activities within a specific period of time. Examples are the Daily Operations Report and the Weekly Operation Summary Report.

Logbooks are used for recording the performance of routine functions. Logbooks are classified according to their application namely, Operations Journal and Technical Logbooks.

An **Operations Journal** is used to record all the guard's activities and observations while on duty. It is used to record the guard's general compliance to the performance of his functions. All guards should maintain a security operations journal while on duty.

The details gathered in the Operations Journal could be used in drawing lessons for the continuous improvement of the security process.

Purposes and Uses of the Security Operations Journal

The guard should know the importance of preparing a Security Operations Journal so as to compel him to properly and meticulously accomplish it. The purpose and uses of the Security Operations Journal are as follows:

1. The journal is used to chronologically record all the guard's activities, performance of functions, observations and instructions received while on duty.

2. It provides for a proof of authority for the guards on duty.

3. It provides for a supplementary source of information by management for policy formulation.

4. It provides for a credible and irrefutable source of information in an investigation.

5. It documents instructions of superiors.

6. It provides for documentary evidence in court if necessary.

Guidelines for Accomplishing a Security Operations Journal

The following are the guidelines on how to accomplish the Security Operations Journal. These guidelines are essential for keeping the integrity of the logbook. The format of the content is illustrated in Appendix 1.

1. All security posts and positions must be provided with a journal.

2. Each guard should have a supplementary personal pocket journal that he can carry with him while on duty. The entries on the supplementary journal should be transferred to the journal at the end of his shift.

3. The journal should be covered and labeled with the document classification, agency name, client name, post, logbook name, and its inclusive dates or period of use.

4. Before using the logbook, the guard must ensure that it has no defect. Examples of these defects are: no page numbers, page misprints and double prints, missing pages, etc. Any defect must be corrected be noted on the logbook.

5. Only indelible ink must be used in accomplishing the logbook.

6. Only universally accepted language, initials, abbreviations, acronyms and contractions shall be used in the logbooks.

7. When accomplishing the logbook, ensure that all the element (when, who, what, where, why and how) of a complete report must be satisfied in all entries.

8. No lines should be skipped in accomplishing the logbook.

9. No page must be skipped or detached from the logbook.

10. All honest errors must be canceled out with a single horizontal line. Errors should remain legible to show that it was in fact an honest error and not an attempt to alter the entry. The correction must be placed on top of the error or immediately after the error.

11. Large unused spaces or pages must be canceled out by writing a big "X". The guard should write his name and sign at the bottom of the space.

12. All signatures used in the logbook must be identified with the hand-printed name of the signatory.

13. The immediate supervisor of the guard or officer must certify the entries in the logbook, at least every shift by writing: "I hereby certify the correctness of all the guard's entries from 8 o'clock A.M. to this time."

14. The Commander must certify the logbook when it is filled up and before it is filed for safekeeping and retention.

15. An index of all unusual incidents entered in the logbook must be made at the last few pages of the logbook for ready reference and to classify which logbook must be retained or disposed of.

16. A system for retention, filing, safekeeping and disposal of all used logbooks must be adopted.

A sample format of the content of a Security Operations Journal is illustrated at Appendix 1.

Technical Logbooks

A Technical Logbook is a record of the compliance of the guard of a specific security function or procedure. A technical logbook generally has a prescribed format for the information solicited and it is accomplished in strict chronological order. Its contents are classified and it is proprietary to the organization. Examples of these are the Visitor's Logbook, Incoming Materials Logbooks and Patrol Checklist.

The guard must record his observations when performing the following functions in their respective logbooks:

VISITOR AND ACCESS CONTROL

Access Control is the process of preventing the intrusion in critical spaces and areas within the facility by unauthorized persons.

Visitor Control, on the other hand, is the process of screening out undesirable or illegitimate persons from entering the facility.

Visitor and access control are also based on the following principles:

1. The uncontrolled entry of individuals in a facility provides

opportunity for undesirable persons to enter, steal and cause damage to resources. Only persons with legitimate purpose shall be allowed inside a facility.

2. Non-organic persons who have been allowed entry into the facility should only be allowed to access areas they have authority to access.

Guidelines for Visitor Control

The following guidelines should be followed to control visitors entering the premises and to accomplish the visitor's logbook either using a hard copy or a digital log:

1. The names of all team members who are authorized to have access to the facility must be submitted to the assigned guard for security purposes.

2. All persons authorized to enter the facility must wear their identification card (ID) at chest level.

3. All persons who are not in the list must have a written authority from the concerned officers of the organization. Such authority must be valid only on the day the subject access is authorized.

4. All persons entering the facility without ID or written authorization are deemed non-organic personnel and should be processed as visitors.

5. The guard is responsible for visitor control. He shall follow the following procedure in processing visitors:

   5.1 He shall cordially greet all visitors.

   5.2 He shall establish the legitimacy of the person requesting entry into the premises.

   5.3 The guard should verify a visitor's appointment by

calling the person to be visited.

When the visitor claims that the person visited knows him/her, a short physical description of the visitor should be verified with the person visited.

After obtaining the clearance from the person visited, the visitor shall be instructed to accomplish the visitor's logbook. The guard shall see to it that all information being asked for is appropriate and complete.

5.4 The visitor's logbook should have adequate spaces for all the following information:

5.4.1 Date

5.4.2 Time in

5.4.3 Time out

5.4.4 Complete name

5.4.5 Complete address or complete name of company with address

5.4.6 Destination

5.4.7 Person to visit

5.4.8 Purpose (specific purpose or activity)

5.4.9 Signature

5.4.10 Remarks

6. The guard shall ask for a valid and authentic letter of authority and/or identification card with picture. He shall see to it that the picture matches that of the visitor. He shall also see to it that the information in the ID matches those that are being entered in the visitor's logbook. Any discrepancy shall be verified and corrected. Any unsatisfactory answer shall be further scrutinized and the person visited must be cautioned. The visitor must be the one to write in the logbook.

JOEL JESUS M. SUPAN

7. After ensuring the legitimacy of the visitor, the guard shall issue a visitor's ID. The visitor shall be instructed to pin the ID at the chest area and be apprised of the area he is limited to have access to. He should follow the shortest possible route to the destination and back. The ID of the visitor shall be left with the guard upon exit.

8. The visitor shall also be given a visitor's pass. The visitor's pass shall determine if the visitor has indeed called on the person he register to visit and that he went directly to and from his destination.

9. The visitor's pass shall bear the following information:

    9.1 Name of person visited.
    9.2 Department
    9.3 Purpose
    9.4 Time in
    9.5 Time released by person visited
    9.6 Signature of person visited with his hand printed name.

10. The person visited shall sign the visitor's pass with the time of release and his printed name. The visitor shall present the visitors pass to the guard on his way out.

11. The guard shall ensure that the person visited has actually signed the visitor's pass and that the duration between the time of release and the time of check out is reasonable. Any undue delay should be verified.

12. In case of known dignitaries accompanied by authorities, no scrutiny is necessary. The guard shall however accomplish the visitor's logbook for the guest. The visitor's ID must however be given to the guest for him to wear.

13. At the end of office hours, the guard must account for all visitors. All unclaimed IDs must be verified from the person being visited. If the visitor has already been released, the

guard shall inform the patrol guard to locate the visitor. Only when patrol is sure that the visitor has left that the alarm will be lifted. Appropriate report of the incident must be entered in the journal, remarks and daily operations report.

Sample form of the content of the logbooks for visitors is illustrated at Appendix 3.

MATERIAL MOVEMENT MONITORING

The members of the procurement team are the ones who are basically authorized to perform the movement and control of materials coming in and out of the facility. The guard's task is to monitor the movement to provide a redundant security measure because materials are generally susceptible to loss.

How to Perform Material Control and Accomplish the Materials Logbook

Principles of Property and Material Control Procedure

1. No one is allowed to bring company property in or out of the facility without a written authority from concerned officers.

2. The guard must record all materials and correspondence being brought in or out of the facility. This record will serve as a back-up measure for existing material receiving procedures.

Purpose of the Material Logbooks

1. The material control procedure is intended to ensure that all company properties and material being brought in or out of the facility are authorized and properly documented by concerned authorities.

2. The material logbook is meant to be an independent and redundant back-up documentation of all materials leaving or

entering the premises.

3.   It is designed to identify the person responsible for the property or material being brought in or out of the premises.

Guidelines and Procedures in Accomplishing the Incoming and Outgoing Materials Logbook

1.   All material and properties being brought inside the facility shall be recorded in the incoming material logbook provided for by the client.

2.   The Incoming Material Logbook shall have the following information on the material:

   2.1   Date of delivery
   2.2   Time of delivery
   2.3   Quantity of observed items
   2.4   Unit or configuration of observed items
   2.5   Particulars or description of the observed items
   2.6   Source of the items
   2.7   Number of the delivery document
   2.8   Name of the courier
   2.9   Person who officially received the delivery
   2.10   Any applicable remark

3.   The Outgoing Material Logbook shall have the following information on the material:

   3.1   Date of delivery
   3.2   Time of delivery
   3.3   Quantity of observed items
   3.4   Unit or configuration of observed items
   3.5   Particular or description of the observed items
   3.6   Destination of the items
   3.7   Number of the delivery document
   3.8   Name of the courier

    3.9   Person who authorized the release of the goods

    3.10   Any applicable remark

4. The guard should not open any sealed receptacle.

5. The guard should only document what he personally observed and not copy what is written in the document presented.

6. The guard shall not receive any package from a questionable source but should refer the matter to his supervisor.

7. The guard shall not receive package of questionable or suspicious features or descriptions such as wrong addressee, wrong or vague address, signs of tampering, inconsistent documentation against what is actually observed, unusual odor, unusual discoloration, leakage, oily or powdery substance, disproportional weight, unusual clatter or sound. He shall report such observation to his supervisor and effect emergency procedure.

8. The signature and identity of the person who actually and officially received the items must be verified from the delivery receipt and from the person himself and entered into the logbook.

9. The guard must obtain and secure a specimen signature of all persons authorized to receive and release materials or properties from the office premises. The specimen must be original and should include the full signature, abbreviated signature and the hand printed name of the signatory.

10. The guard shall at all times compare the actual signature in the property pass to the specimen signature when logging out items.

    Samples of the logbook contents for the incoming and outgoing materials are at Appendices 4 and 5.

JOEL JESUS M. SUPAN

REPORTING

Reporting is an accounting of what has been observed while the guard is on duty and passing it on to another person. It is done orally for timely dissemination and in writing to preserve the accuracy of the information on the circumstances surrounding an activity or an event.

There are basically two types of security reports. These are the routine and non-routine reports. Routine reports are the guard's report of his routine activities and the performance of his functions. Examples of these reports are the Daily Security Operations Report or the weekly operations report, which is a summary of the daily operations report.

Non-routine reports are those reports accomplished on the basis of need and urgency. Incident Reports and Investigations Reports are examples of this type of security report.

A written report is a part of recording. Regardless of the type, reports must have the following qualities:

1. **Accurate**. An accurate report gives the factual and precise information as they happened and as obtained. This can be obtained through records or first hand information. While first hand information leaves a room for doubt, supplemental first hand information from an independent source would provide the required accuracy.

2. **Complete**. A report must provide all the answers to all the essential elements of information and other supporting information. The essential questions are those whose answers would constitute the framework of the picture. The supporting pieces of information are those which give a more vivid description of the event.

3. **Concise**. A report must be straight to the point and must

present only the most essential description.

4. **Organized**. A report must conform to the conventions of proper composition as much as the essential elements of information are presented logically and systematically.

5. **Understandable**. The report above all must be understandable not only to the person concerned but all those who might read the report. This means that it must convey the true picture, which the reporter is trying to draw.

6. **Neat**. A report must be clean, presentable and in proper format. This is a manifestation of respect to the reader and respectability of the writer.

7. **Timely**. A report must be submitted as soon as possible to the concerned so that appropriate decisions and actions based on knowledge can be done in response to the incident.

Elements of a Complete Report

The essential elements of a complete report answers the essential questions, what, who, where, when, why and how:

1. **What**. The answer to this question can be any of the following:

   1.1 A situation that may cause harm, loss, or damage to person and properties as observed by the author of the report.

   1.2 An event where harm, loss, or damage has been caused as observed by the author.

   1.3 A situation where harm, loss, or damage resulted to a recent past occurrence upon a person or property as observed by the author of the report

2. **Who**. The answer to this question can be the persons present

in an unusual situation or event or persons who where involved in an incident either as the perpetrator, the victim, witness or responder.

3. **Where**. The answer to this question can be the places where the unusual situation is present or the event occurred. It also includes subsequent places where persons have gone or objects have been brought to or where the witnesses were situated during the incident.

4. **When**. The answer to this question may refer to the time or instance when the unusual situation has been observed or the event has occurred. This also includes time of responses and subsequent events.

5. **How**. This refers to the sequence of events leading to the observation of an unusual situation. It is also the sequence of actions involved during the event itself.

6. **Why**. The answers to this question are the causes of a situation or of an event as actually observed by the witness or by the author of the report. Often the answers to this question cannot be obtained by mere observation. In this cases the need for the answers become the objectives of an ensuing investigation.

Samples of an incident checklist and a narrative report are illustrated with Appendices 106 and 107.

GUARD COMMUNICATION

Proper Use of Telephones

The guard that is stationary to his post must have a telephone as his primary means of communications with the rest of the team members inside and outside the facility. Oftentimes, he also acts as the telephone operator that initially receives calls from outside. In such cases, he would be in fact, the first person that outsiders will talk to. His demeanor during the interaction would reflect the image of the organization as a whole.

Because of this, it is imperative that the guard must be proficient in carrying out a conversation over the telephone. The following are the guidelines in handling communications through the telephone.

1. Always use a "voice with a smile", much like a warm handshake.

2. Answer calls promptly after the first ring if possible.

3. Identify yourself. If you are the caller, give your name and that of your company. If you were the party being called, state your name and department.

   3.1 Example for incoming call: " Thank you for calling (client's name). This is guard (guard's name), Good (Morning), How may I help you?"

   3.2 Example for outgoing call: " Good (Morning), this is guard (guard's name) of (unit). May I speak with (name of the subject person)? Thank you.

4. Keep a pad and pen handy. Jot down details of messages accurately.

5. When the person being called is not around, give the caller a definite time when to call again. Offer to take the messages.

6. When screening calls or asking for the identity of the caller, avoid using "Whose calling?" or "Who is this?" "May I know who's calling?" sounds more polite.

7. Be polite when you get a wrong number call. Instead of shouting "Wrong number!" You might say, "I'm sorry, you dialed the wrong number".

8. Finally, end the conversation politely. Do not forget your "Thank you" and "Good bye".

JOEL JESUS M. SUPAN

Proper Use of Radio

The fastest, most reliable and convenient means of communication of the guards on duty is the two-way radio. The following are the guidelines on the proper use of the radio:

1. Do not use call signs and coded terms. Call signs and coded terms are more appropriate in military or police operation to prevent disclosure of identities. But they are deemed impolite in a corporate environment. Use your name, the name of persons and post location as they are called.

2. Talk only when rendering reports, acknowledging, or giving instructions.

3. Do not transmit classified information over the radio.

4. Do not play with the radio. It is discourteous.

5. Do not monitor or report locations of executives over the radio except when asked.

6. Do not turn on the volume so much as to disturb the people around you or hear your conversations

VEHICLE CONTROL AND INSPECTION POLICY PROCEDURE

Principles of Vehicle Inspection

1. All proprietary vehicles must be in good running condition at all times.

2. Any change that can be perceived by the senses on all proprietary vehicles shall be properly documented.

3. Vehicles shall be used only for official and explicitly authorized purposes.

Objectives of Vehicle Inspection

1. To ensure that all company vehicles are in good running condition at all times.

2. To ensure that only authorized person with official and authorized purpose shall be allowed to use company vehicles.

Guidelines and Procedures for Vehicle Inspection

1. Using the vehicle inspection checklist the guard shall inspect all company vehicles before office hours, before use and after use.

2. Before departure, use, and after arrival, the guard shall assist the authorized driver in checking the brakes, lights, oil, water, battery, air, tire, shifting, gas, wipers, ignition and mirrors and document all observations.

3. The guard shall ensure that all vehicle doors are closed if subject vehicles are unattended.

4. The guard shall include all vehicles in his routine patrol.

5. When the guard is assigned in a public parking lot, he must maintain a Vehicle Monitoring Checklist and record all vehicles parked in the facility. The checklist should capture the type, make, model, color, plate number, and external parts such as wipers, logos, tires and defects.

See Appendix 13 for an illustration of a sample the Vehicle Inspection checklist.

KEY CONTROL PROCEDURE

Key Control is the function of whoever is in charge of the physical facility. But often the execution of procedure is delegated to the guard. Next to the guard, the key is the most popular symbol of security.

JOEL JESUS M. SUPAN

The key is the primary instrument used to secure spaces and receptacles from unauthorized intrusion. Security is said to be compromised when the key falls into the wrong hands. A key that is lost but does not fall in the wrong hand could result to obstruction that would impede the operation of a routine or emergency function of the organization. These are the reasons why a guard should know Key Control Procedures.

The Objectives of Key Control

1. To ensure that unauthorized persons shall not gain access to restricted areas, inclusions, classified equipment, materials, and information.

2. To ensure expeditious access to a particular space or locker whenever necessary.

Guidelines for Key Control

The following are the guidelines for an effective key management and control.

1. The guard shall be charged with the custody and responsibility of all inclusions, restricted areas and lockers.

2. All keys shall be provided with a key control card. The key control card shall have the following information:

   2.1   Key number
   2.2   Lock name or location
   2.3   Date acquired and date installed
   2.4   Number of duplicates upon acquisition
   2.5   Number of duplicates with respective dates of duplication
   2.6   Distribution of duplicates
   2.7   Name of manufacturer, duplicator, address and telephone numbers
   2.8   Relocation of lock, if any

2.9   Proof of receipt by person in custody of a key

3. All key cards shall be contained in a secured key catalogue.

4. All original and excess duplicates shall be stored in a secured key cabinet.

5. All key identification tags shall be coded. A decoding table shall be in the custody of the Master Key Custodian.

6. The Master Key custodian shall maintain a key borrower's logbook. The logbook shall have the following information:

    6.1   Name of borrower
    6.2   Purpose
    6.3   By whose authority
    6.4   Date and time borrowed
    6.5   Date and time returned
    6.6   Name of issuer
    6.7   Declaration that no duplication was made

7. All keys at the key locker must be accounted for every end of the office hours.

8. Any lost key must be reported immediately.

9. Locks of lost keys must be reconstructed or relocated as soon as possible.

10. All unsecured locks of restricted areas and inclusions must be repaired.

11. All keys must be embossed with identification marks.

12. Any key that is found unattended must be surrendered to the guard or to the Master Key Custodian.

13. A key and locker inventory and showdown must be done every six months.

A sample of a Key Control Card and a Key Borrowers' Log is attach at the end of this Section as Appendices 14 and 15.

JOEL JESUS M. SUPAN

PUBLIC RELATIONS

One of the most critical functions of a guard is public relations. This is because it is the public that will rate the performance of the guard. The rating will not be based on efficiency and effectiveness alone but also by how he has left a positive impression of professionalism to the people he has dealt with.

A good public relation is a condition where an individual or entity has an understanding and goodwill of the people around him. Achieving expectations can foster good public relations.

To achieve this, it is essential to know and understand the different classifications of people that the guard normally deals with. They are the following:

1. The management
2. The team members
3. The management agency if the guard is contracted
4. The guard's supervisor
5. The government representatives
6. The guard's co-guards
7. The general public (client's clients and visitors)
8. The violator

The basic expectations of the public of the guard are as follows:

1. The guard must look like a professional
2. The guard must speak like a professional
3. The guard should perform like a professional

The guard can look like a professional by having or performing the following:

1. Proper grooming
2. Complete, neat, and presentable prescribed uniform
3. Complete, clean, and ready to use operational equipment and accessories

The guard is said to be speaking like a professional if he manifests courtesy in his speech to every one he deals with. Thus,

1. He must address everyone as either Sir, Ma'am or Miss

2. He must say "please" to solicit the cooperation of persons or when asking favors or giving directions.

3. He must say, "Excuse me" when giving caution or correcting the misdeed of another.

4. He must say "Thank you" to everyone who cooperated, recognized of acknowledged his request.

The guard is said to perform like a professional if he does the following:

1. He has mastery of the basic functions of the guard by knowing and expressing by heart his function, by knowing how to perform them and why he performs them.

2. He is firm in his enforcement of regulations; he is fair, friendly and performs his functions with dispatch.

REGULATION ENFORCEMENT

One of the most difficult functions of the guard is Regulation Enforcement. This is because it entails the guard's correction of the misdeed of others. This, however difficult, can be addressed effectively by following these guidelines:

1. The guard must have a copy of the regulations at all times. This strengthens the position of the guard and it demonstrates that the guard is just under order from a higher authority. With it, the guard can correctly site the specific regulation that he is enforcing.

2. The guard must apprise team members of the regulation as applicable even before they commit violation. This can

prevent the member from committing the offense at the outset. It is harder to correct a misdeed than to prevent it since the former causes a dent on ones ego.

3. The guard must know the steps for all regulatory procedures. He should be able to guide the members on how they should perform or behave.

4. The guard must ensure the availability of an instrument or support to enforce the regulation. The guard must have the proper and appropriate forms for recording when asking for information. He must also have the proper physical restraint or delineators to physically guide people.

5. The guard must be firm, fair, friendly and fast in enforcing regulation.

REACTING TO EMERGENCY

One of the most stressful functions of the guard is Reacting to Emergencies. This is because this function involves risking his personal safety. The guard must base his reactions on the principles of reacting to emergency as follows:

1. Assess the general safety of the situation before reacting.
2. Assess the risk if it can be subdued or controlled. If so, implement the prescribed control measures.
3. Call for emergency assistance.
4. Secure the area until assistance arrives.

Guidelines for Reacting to an Emergency

Emergency Telephone Numbers. All team members having with them communication equipment, such as telephone and radio, must have with them all the time the following emergency telephone numbers:

1. Nearest police station

2. Nearest fire station
3. Nearest hospital
4. Nearest bomb disposal unit
5. Nearest poison control
6. Energy supplier
7. Water utility
8. Building security
9. Security agency
10. Key company officers

The guard must also have an alternative number or alternative emergency facility to call in case the primary emergency facility is not accessible.

The guard must conduct a daily line check of these numbers to ensure that they are operational during his shift. He shall disseminate any changes to the operator and his reliever, as necessary.

REACTING TO CRIMINAL INCIDENT

Guidelines for Reacting to a Criminal Incident

1. In any event of criminal activity, the Guard should first check if the perpetrator were armed and that he had adequate means to subdue him.

2. In all cases, the assistance of the police must be sought for. Armed assailants must not be confronted nor engaged. It would endanger the lives of others or even the guard himself.

3. The guard must obey all the orders of the assailant so as not to provoke any confrontation.

4. The guard should not make sudden moves to provoke the assailant. Permission must be asked before making a move.

5. Observe the assailant for any outstanding characteristics, such as birthmarks, tattoos, deformities, etc. The guard must

observe the general characteristic such as height, built, sex, age, complexion and race. Then, he should observe any other specific characteristic or feature such as hair, eyes, ears, nose, mouth, chin, neck arms feet and voice, its accent, rate and pitch. He must remember the exact words of the assailant.

6.   After the incident, the guard must preserve the area, not letting any one leave or enter the area until the police take over the situation.

Making a Citizen's Arrest

The law allows a citizen's arrest, which is also called warrant-less arrest, of identified criminals. However, there are requisites by which this is performed. They are as follows:

1.   When a crime is being committed or about to be committed in one's presence;

2.   When there is a compelling reason to believe that a crime has just been committed;

3.   When the suspect is an escaped convict.

The Principles of Arrest

A guard is said to be renege in performing his duties if he saw a crime being committed or other requisites for a citizen's arrest were present but did not make the arrest. However, he has to take into consideration the principles of arrest to guide him on how to effectively arrest and at the same time do not compromise his and other's safety. The principles of Arrest are as follows:

1.   Perform an arrest when a crime is perceived to be committed and all the requisites of a citizen's arrest are present.

2.   Ensure superiority of strength to subdue the suspect. If the suspect is much superior, call for back up.

3.   Apply just enough force to reasonably subdue the person being arrested.

4. Read the rights of the suspect.

5. Use reasonable means of restraint.

6. Apply appropriate prescribed period of lawful detention.

Use of Force

If the presence of the guard alone is not enough to deter illegal acts and if the threats are high, the guard may have to use force to protect people and properties. The following are the guidelines that the guard must follow if he had to use force:

1. The guard must respect human rights and uphold the dignity of all persons.

2. The guard must first try to use non-violent means and use force only when necessary.

3. The guard must use only the minimum force required. He should keep the force proportional to the threat. The following are succession of actions by degree of force:

    3.1 Use voice in making a request, an order and admonition;

    3.2. Use the whistle;

    3.3 Use hands for defensive tactics and restraint;

    3.4. Use the nightstick for defense and for restraint;

    3.5 Use the service firearm means of restraint have been exhausted without success and if all the elements of self defense were present.

4. The guard must always act within the bounds of the law and the authority given when using force.

5. The guard must minimize injury and damage.

6. The guard must give medical aid or assistance to injured persons, including offenders.

7. The guard must report any use of force as soon as possible his supervisor.

8. The guard must inform the relatives of any persons injured or killed by the use of force.

The Elements of Self Defense

There could be instances when the guard has to confront an armed assailant in self defense. Self defense as defined by various laws can only be invoked when all the following elements are present:

1. When there is unlawful aggression against the guard;

2. When there is necessity of the means employed to defend another;

3. When there is lack of undue provocation on the part of the guard.

The Elements of a Lawful Defense of Other Persons

The guard may also be put in a situation where he has to defend the life of others. This can be done only when all the following elements are present:

1. When there is unlawful aggression against another person.

2. When there is necessity of the means employed to defend ones self;

3. When there is lack of undue provocation on the part of the guard or the person being defended.

HOW TO REACT TO BOMB THREAT

The mere presence of a bomb in the facility constitute a bomb threat, it must be addressed immediately to prevent any untoward incident. This is the reason why it is imperative that a guard or the other team members should know how to identify an explosive devise. The guidelines for the identification of bomb or improvised explosive devices are as follows:

1. The guard should always be on the look out for objects that do not belong to the place.

2. The guard must be on the look out for bomb components either bundled together in a single contraption or placed apart from each other in and among the baggage of a single or several persons. The components of an improvised explosive devise are the switch, power source such as a battery, initiator, the containment or compactor and the explosive.

3. The guard should always be on the look out for unattended packages, bags or other receptacles.

4. If a suspicious package is observed, do not touch it. Call the supervisor at once, who in turn should call the bomb disposal unit. Implement the procedure on emergency action for bombs and let the police do the rest of the procedures on bomb threat situations.

5. Watch for people with unusual, demeanor, walk, disposition or facial expression.

How to React to Bomb Threats Over the Telephone

Bomb threats given over the phone are in general hoaxes. However, they cannot be set aside as such and has to be verified as soon as possible. The best way to verify this is from the caller himself. A guard or any team member must know by heart the correct process on how to respond to bomb threats over the telephone.

1. All telephone sets must have a bomb threat guideline beside them. The guidelines must provide for the things a receiver

must ask for. A sample checklist is at Appendix 8.

2. Anybody who receives the call must attempt to prolong the conversation. He must not put down the phone but exhort the caller to talk.

3. He must take note of the time of the call and signal silently to the nearest person about the call for the latter to inform concerned officers.

4. He must elicit the following information by order of urgency:

    4.1 What time that the bomb would go off?

    4.2 Where the bomb was placed?

    4.3 How does the bomb look like?

    4.4 How is the bomb activated?

    4.5 What type of bomb is it?

    4.6 Why is he doing it?

    4.7 Does he know that people can get hurt?

    4.8 How will he be called?

5. At the same time, the receiver must listen to the background sound such as traffic, airplanes, trains, ships, machines, music, doors and windows closing, and sounds of children or animals.

6. The receiver must take note of the voice characteristics of the caller. Take note of the accent, mannerism, rate, pitch, volume, muffled or signs of alteration.

7. Hold on to the caller as long as possible.

8. Management should decide whether a search for the bomb or for an evacuation to be made. Security must however perform their respective tasks as prescribed by the evacuation policy. The police must be informed of the call in all cases.

How to React to Explosions

A guard must train himself to develop the instinctive reaction to unusual events. One such event is explosion in his area of responsibility while on duty. He should master the following procedures:

1. A bright flash of light always precedes explosions. Immediately duck to the ground or the floor.

2. Crawl away from the path of people who may stampede.

3. After the explosion, attend to the victims, if any. Give priority to those whose injury or condition is life threatening.

4. Look for other witnesses and identify them. Get their contact numbers.

5. Secure the area surrounding the scene of explosion.

6. Contact your supervisors who should report the incident to the authorities.

7. Be ready to make your own statement to the investigators

FIRE SAFETY AND EMERGENCY PREPAREDNESS

Fire is one of the most devastating incidents that can occur to a facility. This incident must be addressed by the Crisis Contingency Team, which includes the fire fighting team. However, the guard is primarily involved in the enforcement of fire prevention and in the initial stage of the fire should he be present at the scene.

The following is the procedure to be performed by the guard for fire prevention and preparedness:

1. The guard must look for any fire hazard during his patrol. He shall clear and correct any observed fire hazard or cause the correction if it is beyond his capability.

2. All team members must report to the guard any fire hazard

they may observe in the office, including other team members
who created the hazard.

3. The guard must conduct daily inspection of all fire alarm
stations and fire fighting equipment and ensure that they are
ready for use. The guard should also monitor the performance
of the team members designated to do inspection and testing
of fire fighting equipment and he should call their attention
if they fail to do so.

4. The guard and all team members shall familiarize themselves
with the building fire defense plan and evacuation procedures.

Emergency Actions in Case of Fire

1. Any one who discovers a fire shall sound the alarm or send
someone to sound the alarm.

2. If the fire has just started, fight the fire with the use of the
nearest portable fire extinguishers.

3. If the fire alarm is set on, call the fire department.

4. If there is a need to evacuate, follow the prescribed evacuation
of the building.

EMERGENCY ACTION FOR SICKNESS AND INJURY

In case any employee, visitor, or other individual incurred
sickness or injury, the guard should do the following activities based
on the Principles for Emergency Action:

1. Assess the situation if it is safe for you to touch or attend to
the victim.

2. Make a primary survey of the victim's condition.

   2.1 Check for responsiveness, if conscious or not.

   2.2 If the victim does not respond, check for blocked
   airway, breathing, and circulation (ABC).

    2.3   If any of the ABC were absent, apply cardiopulmonary resuscitation (CPR).

3. If and when the victim responds, call for emergency help.

    3.1   Call the hospital, doctor and ambulance.

    3.2   Tell them your name location and telephone number.

    3.3   Tell the condition of the patient and the possible cause of the illness.

    3.4   Tell them what help is being done.

4. Make a secondary survey of the victim.

    4.1   Check and monitor the vital signs:

        4.1.1 Rate of breathing, if normal (15 to 20/minute)
        4.1.2 Rate of pulse, if normal (60 to 80/minute)
        4.1.3 Skin, if hot or cold.
        4.1.4 Skin, if moist (clammy) or dry (parched)
        4.1.5 Pupils, if dilated, constricted or uneven.

    4.2   Make a head to toe inspection of the victim. Look for:

        4.2.1   Bumps
        4.2.2   Depressions
        4.2.3   Swelling
        4.2.4   Concussions
        4.2.5   Sore or Painful Spots
        4.2.6   Wounds
        4.2.7   Bleeding
        4.2.8   Discoloration
        4.2.9   Disfigurement/Distortions
        4.2.10 Abnormal secretion or salivation
        4.2.11 Foul smell

    4.3   Apply necessary aid to ease the suffering of the victim until medical assistance arrives.

GUARD MOUNTING PROCEDURE AND TURNOVER OF DUTY

Turn over of guard duties is the most critical phase of the guard's tour of duty. This is because the effectiveness of the incoming guard is dependent on the inputs of the supervisor or the outgoing guard. The incoming guard must be apprised on the situation of the facility from the time he was relieved to the time he assumes his duty.

Principles of Guard Mounting

1. The supervisor shall ensure that all guards are fit and operationally prepared to stand duty.

2. A well-informed guard is a forearmed guard.

3. Continuous education that is done during turnover makes up a complete guard.

4. The motivation and the effectiveness of guards are dependent on the supervisors with whom they interact with during guard mounting.

Guidelines for Guard Mounting

The following are the guidelines for guard mounting:

1. The supervisor must form all incoming guards at least 20 minutes prior to assumption of duty.

2. The supervisors shall conduct attendance check and rank inspection. He shall ensure that all guards are fit to work, not sickly, not drunk, and have no physical or emotional impediment.

3. The supervisor shall inspect the guards individually to check if their uniforms and paraphernalia are complete, proper, prescribed and presentable. The inspection must include the proper way of wearing such.

4. The supervisor shall brief all guards on the following:

    4.1   Latest instruction of client

    4.2   Current security situation

    4.3   Significant development on the peace and order situation in the locality

    4.4   Significant news item for the day

    4.5   Security procedure reminder for the day

    4.6   Client policy reminder for the day

5. The supervisor shall synchronize the time of the timepieces of the guards.

6. The supervisor shall dismiss all incoming guards for deployment to their respective posts.

7. The supervisor shall then form all outgoing guards and conduct a cursory inspection of the guards on their appearance.

8. The supervisor shall solicit and receive debriefing for significant events, observations and instruction by officials, from the guards.

9. The supervisor shall give the basic reminder and the lesson for the day.

10. The supervisor shall dismiss the guards. He shall then make them sign the time cards or sheets, as applicable.

11. The formal way of guard mounting is when the supervisor shall lead the incoming guards to all the post and he supervises the turnover of guards in each post. The incoming guard will be left behind and he will be replaced in the ranks by the outgoing guard. The supervisor shall then dismiss the outgoing guards when all the post turnovers are completed.

12. The supervisor shall ensure that all outgoing guards are out

of the facility within 15 minutes unless there is a valid or authorized reason. A guard who leaves late from the premises shall be treated as a visitor and therefore shall log out in the logbook.

13. The supervisor shall then patrol and inspect all posts to ensure that proper turnover of duties was carried out.

Guard Turn Over of Duties and Responsibilities

Principles of Turn Over of Duties

1. Turn over of function is a critical point to any security guard operation.

2. The guard on duty is solely responsible for any defects in his area of responsibility (AOR) even if such occurred before his duty.

Guidelines for Turnover of Duties

1. The incoming guard shall report at his post at least 15 minutes before posting.

2. The incoming guard shall ensure that he is fit to stand duty. He should be in complete and presentable uniform and properly groomed.

3. The outgoing guard shall make a cursory inspection of the incoming guard for fitness to duty and completeness and presentation of uniform. Defects must be corrected.

4. The outgoing guard shall log in the actual arrival of the incoming guard before turning over his post.

5. The incoming and the outgoing guard shall make an inventory and inspection of all equipment, logbooks and tools required in the course of the guard's duty. The inspection shall cover completeness, cleanliness, orderliness, and readiness for use

and operation. The guards shall log in all their findings of the equipment.

6. The incoming and outgoing guards shall conduct a patrol of the post and the area of responsibility. They shall use the patrol checklist for this purpose. The patrol checklist shall cover all area control points and critical points.

7. The incoming guard shall read all logbook entries from the time he was relieved to the time before re-assuming duties. This is so he knows what happened in the post during his time-off. He shall certify on the operations journal that he has read all subject entries.

8. The guard shall assume duty and log in the actual time of assumption.

9. The guard shall log the actual time of departure of the outgoing guard.

Summary

The guard is the most prolific of all security measures available to the team. The guard is an extension of the eyes and ears of the leaders. They have routine access to places in the facility which the leaders do not usually go to. They provided the needed information on the facility. They provided redundancy for the inherent controls set up by the leadership for the protection of the teams resources

-oOo-

CHAPTER 9

Security Guard Force Management

There are basically two types of Guard Forces as to ownership and management. They are as follows:

1. **Proprietary Guard Force**. This a guard force that is organized by the company, where the guards are deemed as employees. As such, they are given the same benefits of any other employee. This type applies to organizations and government agencies in some countries.

2. **Contracted Guard Force**. The guards are contracted through a security agency or company that is authorized and supervised by the government.

Both types of services have their respective advantages and disadvantages.

The members of the proprietary guard forces are regular employees. They enjoy the same benefits and compensation on package of other employees in the organization. One distinct advantage of this type of guard force is that they develop loyalty to the company and their familiarity of the organizations and its facilities. This is necessary for effective performance of duties. They likewise report administratively and operationally to one supervisor.

A distinct disadvantage of a proprietary guard force is that the guards have little chance for career development compared to the other line employees because of their specialized functions. It is difficult to move them from one department to another in case of misconduct. Because of their tenure, the cost of maintaining them becomes more and more expensive. Moreover, their continued employment in some instances, breeds undue familiarity with the employees that could impede an effective enforcement of regulation, which is one of the basic functions of a guard.

On the other hand, contracted guards are less costly. Moreover, the organization is shielded from extraneous expenses that it would otherwise bear if they were regular employees. The company can also change the contracted guards or the security agency that provides them should they fall short of the organization's expectations without the complications bordering on labor relationship issues in some countries. One distinct disadvantage of this however, is that they would always remain to be detached from the rest of the employees, thus precluding the development of team bonding that is important in achieving the organization's goals.

GUARD FORCE LEADERSHIP

Whether a guard force is proprietary or contractual, it is essential that a full time member is assigned to manage and lead the Security Force. This is to provide focus to the mission of the security team. This mission should be aligned with the organization's objectives.

Most organizations have a security department lead by a Team Leader. But more often than not, the team leader is relegated to manage physical aspect of security including the guard force, which is akin to the police force in the organization. It is advisable that the security team leader must also be given the authority to be involved in policy formulation and be given the same status as the officers leading other operating units.

This officer must be equipped with the broad knowledge and understanding of the organizational structure and operations. He must have entrepreneurial aptitude. He must have intensive education and extensive experience in organizational management as well as in the security field. He must be competent and proficient in the application of the basic tools of the profession. He must also have a wide network of contacts in the government, in his own field of expertise and the other fields that support the organization.

The Guard Force itself must be lead by a Specialist Security Supervisor who would provide the operational supervision of the guards. The supervisor must also have extensive education and training in security and guard force management.

TYPES OF SECURITY UNITS

Guard Forces are classified into different types of security units. The types of unit are based on the number of guards or posts.

It is important to classify Security units so that the qualification, career development, rank and compensation of those who lead and manage the security team can be determined. There are still no known best practices, let alone a standard to classify security units.

But before this is discussed, it is important to know the terms to be used in this topic. The guard's basic tour of duty in one day is the same as that on the ordinary team member. An eight-hour duty is generally adopted in accordance with the labor laws prevailing in most countries. The tour of duty of a guard for the day is called a "shift". Thus, there are three basic shifts of security guards in one day for twenty-four hours of security guard coverage.

Another term that is basically used for guard duty is the "security post". A security post is the specific place of duty in the facility where the guard is posted. A specific place of duty is called a fixed security post. Examples of fixed posts are the gates, doors, main entrance lobby, loading bay, CCTV monitoring room or perimeter watchtower.

The other types of posts are the designated duties that do not have a specifically designated post but rather, the guards have to move in and around the facility. This is commonly called the Patrol Duty. The following are the types of security units.

1. **Single Post – Single Shift**. This is a type of security post

which requires only one guard. It is generally applied to small businesses that do not operate for 24 hours. Often, the facility has alternate physical security outside of office hours. This type of post is also used for residences to provide watch at night when all the household members are asleep. The guard here is his own supervisor. He reports directly to the team member designated to manage him. The guard here cannot be made responsible to issues that arise outside of his tour of duty unless proven that the cause of the incident is directly attributed to hi negligence during his tour of duty. Often, this guard is also made to conduct patrol within the facility.

2. **Single Post – Multiple Shifts.** This is a type of security unit where there is more than one guard and more than one shift. This type of security unit is generally applied to organizations with twenty-four hours operation and who have no alternative security measures outside of the office hours. This type of unit is also used for residences where the household members deem their need for twenty-four hours security and that they can well afford it.

This security unit would need a senior guard to perform administrative functions to collate administrative requirement for submission to the contracting agency and to receive orders from the client. Operational orders in this type of unit are disseminated during turnover of duties in between shifts.

3. **Multiple Posts – Single Shift.** This is a type of security unit is where more than one guard is needed to be assigned in different posts during a single shift. This is applicable for large facilities with wider areas to cover but, its operation is only as long as a single shift and that they have alternative physical security measure outside of the operating hours. A

bank branch is a typical example of this type.

A head guard is necessary for this type of unit. He shall be performing administrative and operational supervision over the other guards.

4. **Multiple Posts – Multiple Shifts – Single Location.** This type of security guard unit is be called a Security Detachment. It can be a small detachment, medium sized, large or ultra large detachment. The determination of the size is relative to the size of the facility of the client and that it is left to the client to decide the strength of the unit. The strength of the security unit of this type may range from at least nine guards and to an undetermined number. This unit is headed and supervised by a Detachment Commander. He is assisted by an Assistant Commander and/or the Shift Supervisor.

5. **Multiple Posts – Multiple Shifts – Multiple Locations.** This type of a security unit is usually applied in very large organizations with facilities in different locations. Examples are a chain of malls and department stores or multi-national organization. And Over-all Security Commander is needed to supervise this unit. He can be assisted by the Detachment Commander in the different facilities.

GUIDELINES FOR SELECTING A SUITABLE AND RESPONSIVE SECURITY GUARD FORCE PROVIDER

The parameters for selecting between a proprietary guard force and that of the contracted guard force are basically the same. However, there are fundamental differences on the structures and the nature of the business of the two types. Often, the difference is the cost. This is when a contracting a third party security services is favored. Thus, it is imperative that an organization establishes the criteria for selecting the ideal and the most suitable agency. This is to ensure value for the investment that the organization is putting on security.

To be able to acquire the most suitable and responsive Security guard force provider, the following process and attributes should be considered.

1. Check the trade record of the prospective service provider. The record includes the License to Operate, business permits and all legal papers and clearances mandated by the national and local governments. Those providers whose papers are without discrepancies or irregularities should be favored from those with discrepancies and irregularities.

2. Look for the explicit service commitment of the proponent and the plan by which the said commitment shall be delivered. The ultimate measure of a service delivered is that there is no loss or damage to life, property or reputation at a reasonable cost.

3. Should a bidding be opted, the proponents should be made to present and defend an overall integrated security plan utilizing every aspect and element of security and presenting a guard force management plan. Note: This should be an integral part of the subsequent service contract to be executed.

4. The foregoing is based on the premise that the guard services rates per guard are prescribed and the prevailing condition results to cut throat competition by the guard service providers. For this reason, the proponent would be biding with the same cost. A proponent giving a lower price would most likely scrimp on its service if not fail to give the lawful compensation and benefits of a guard. The idea of an integrated security plan is to determine if the proponent can give the same effective security services with lesser guards and therefore, for a lesser cost. The integrated security plan should not be limited to guards alone but also for other high technology equipment, which will enhance the efficiency

of the guards and effectiveness of the system to deliver the desired result. Then, the cost may drastically be reduced and it will be equitable and feasible to the winner.

The following procedure is suggested on how to select the most suitable and responsive Agency:

1. A selection board shall be created and it will lay down the criteria for selection. Examples of the criteria are the following:

    1.1 Cost
    1.2 Over all security plan
    1.3 Reporting systems
    1.4 Guard deployment scheme and thoroughness of specific functions
    1.5 Equipment deployment and utilization plan
    1.6 Frequency of patrol
    1.7 Operational readiness of the present guards
    1.8 Mastery of the Agency Management to formulate Security Programs
    1.9 Reaction time to emergency
    1.10 Investigation capability and skill
    1.11 Office management
    1.12 Compliance to pertinent laws and regulations

2. Values or weights shall be assigned to each criterion according to its significance to the security requirement of the soliciting company.

    Note: The cost to be considered shall be based on the total cost of service rather than on the rate of the unit's service (8 hours per day) for guards. The cost may include subscription for the use of equipment offered for installation to supplant any reduction of guards from what is previously prescribed.

3. The proponents shall each make an oral presentation of their

proposed security plan in the presence of the other bidders.

4. The members of the bidding committee shall rate every bidder based on the particular criterion, say on the scale of 1-10.

5. The rating will be multiplied to the designated weight of the said criterion to get the criterion rating.

6. Then, all ratings of all the criteria will be added to get the total rating of the each proponent.

7. The proponent with the highest total rating will be awarded the contract.

Sometimes the biggest agencies may not be the ideal agency and sometimes even a new and dynamic agency will best serve the security needs of a facility. The plan presented shall then be made an integral part of the contract. Thereafter, all that is needed is to monitor the said compliance. A Checklist to Determine a Suitable Agency is illustrated at Appendix 10.

ELEMENTS OF A SECURITY SERVICE CONTRACT

The Security Guard Service Contract is a formal instrument that provides for the terms and condition of the services and considerations exchanged between the Organization and the Security Guard Service provider. The elements are listed and described as follows:

1. **Title**. This element cites the name of the Instrument and the nature of the services being acquired.

2. **Preamble**. This element provides for the legal identities of the parties to the contract, their respective circumstances and representations.

3. **Obligations of the Service Provider.** This element states the services to be provided by the service provider and how such are to be carried and delivered. The details of this

element should be attached as appendices as follows:

3.1 **Appendix 1** - Post Description and Guard Deployment Schedule;

3.2 **Appendix 2** – Guard Forces Management Plan includes the Organization and Organizational Relationships, Security Operating Policies and Procedures, the Post Duties, Post Responsibilities and Reportorial Requirements;

3.3 **Appendix 3** – Service Level Agreement which summarizes and tabulates all the expectations of the organization from the service provider

4. **Penalty Clause**. This element states the agreed penalties to be meted to the service provider should the expectations of the client organization are not met.

5. **Confidentiality Clause**. This element describes the need to maintain the confidentiality of the contents of the contract, which are to be confined only to the parties and their assigns.

6. **Considerations Clause**. This element describes the cost of the services broken down in detail, the terms and the schedules by which the services are billed and paid.

7. **Exemption Clause**. This element describes the circumstances by which the Service provider is exempted from liabilities such as Force Majeure.

8. **Duration and Expiry**. This element states the life of the contract, which should start upon signing or from the deployment of guards and ends at the expiry date. It also provides the procedure for the renewal or non-renewal of the contract after its expiry. There are practices where the service contract is "evergreen". This contract has no definite life provision. It can however be terminated any time after

due notice. This contract is based on the principle that it is very tedious and very expensive to do bidding periodically and regularly when the effectiveness of the services is based basically on the individual guard performance. The training of new guard will also compromise the continuity of the security of the facility. An erring guard would be sanctioned but such offense is not attributed to the agency. This is most suitable where the security guard provider is deemed professional.

9. **Governing Laws and Settlement of Dispute**. This element provides for the procedures to be followed in the event of a dispute in the implementation of the contract.

10. **Signature and Subscription**. This provides for the testimonies of the representatives of the parties to the contract as to their understanding and acceptance of the provisions of the contract.

11. **Notarial Registration**. This element provides for third party recognition of the contract as a legal and public document.

THE ELEMENTS OF THE GUARD FORCE MANAGEMENT AND OPERATING PLAN AND PROCEDURE

The Guard Forces Management and Operating Plan and Procedure are the basic references of all guards in the performance of their duties. Its provision shall be reprinted and distributed to all the guards and other persons concerned with security. It should be covered by an implementing policy so that appropriate control can be adopted to enforce it.

The elements of the guard force management and operating plan and their descriptions are as follows:

1. **Reference**. The reference cites the justification for the need

of a plan. The basic reference in this example is the Guard Service Contract.

2. **Objectives, Purposes and Uses**. This element describes the rational objectives and the principles of the Guard Force Management and Operations Procedure.

3. **Definition of Terms**. This element provides definitions for the terms peculiar to the described services. It also defines the contexts for which the terms are used.

4. **Statement of Security Guard Service Principles**. This element describes the principles such as the detail on the practical expectations and limitations of the guards. These are based on the accepted principle of guarding that could exempt them from undue liability

5. **Organizational Relationship**. This element describes the line of authority for both administration and operation. By the nature of the service contract in relation to the labor laws in most countries, the contractual guards are not employees of the organization as such they are still under the administrative control of their agency. However for operational expediency, the team leader of the Security Force gets direct instructions from the representative of the organization who contracted them.

The context of operation by a Security Guard Provider should be understood as providing administrative support for the different security teams deployed in the client's premises. This is different from security operations that is the actual performance of security functions by the Security Team and its member guards at their respective Areas of Responsibilities as defined by the client organization.

6. **Deployment Schedule**. This element states the different post and the shifts to be manned by security guards.

7. **Security Requirement**. This element describes the security needs of the organization. They provide the rationale for the strength, deployment and functions to be performed by the guards as individuals and as a team. These requirements are discussed in detail in Chapter 1.

8. **Security Operation Requirements**. This element enumerates the different Security Control Measures and Procedures, which the guards should perform to address the security requirements of the Organization.

9. **Duties and Responsibilities**. This element states and describes the individual duties and responsibilities of the guard in the different post. This duties and responsibilities are based on the Operational Requirements and the Basic Functions of the Guard. The Duties and Responsibilities also describe the organizational relationships of the guard on a particular post, the required operating equipment and the prescribed reports to be accomplished and the schedule of their submission.

10. **Revisions, Amendments and Additions**. This element provides for the latitude to make adjustments to the plan as deemed necessary and as demanded by the circumstances.

11. **Appendices**. This element refers to the attachment to the Plan. It includes the different Security Policies and procedures of the organization, which the guards must enforce. They are (but not limited to) the following:

11.1 Security Patrol Procedure

11.2 Visitor Control Policy Procedure

11.3 Incoming and Outgoing Material Control Procedure

11.4 Key Control Procedure

11.5 Policy Procedure for Vehicle Control

11.6 Emergency Procedures

11.7 Reporting Procedure

11.8 Recording Procedure

11.9 Duty Turn-over Procedure

DUTIES AND RESPONSIBILITIES OF A TYPICAL SECURITY SUPERVISOR

The Security Supervisor is the leader of the security guard force. He is the primary link between the Guard Service Provider and the Client Organization. As such he has to perform the following functions:

The Security Guard Supervisor's Administrative Duties and Responsibilities are as follows:

1. Prepares the periodic Guard Detail;

2. Ensures the availability of supplies and equipment by making timely requisition;

3. Checks the attendance of the guard and accomplishes the Daily Time Record;

4. Accomplishes and submits the daily, weekly and semi-monthly security operations reports;

5. Checks the uniforms of all guards for correctness, completeness and presentability;

6. Checks the fitness of all guards to perform their duties;

7. Maintains the record of all operation and administrative documents of the detachment;

8. Conducts investigation and adjudication of guard misconduct.

The Security Guard Supervisor's Administrative Duties and Responsibilities are as follows:

1. Conducts mustering, inspection and briefing of the guards prior posting using the prescribed procedure for guard mounting;

2. Inspects the guards on post during their tour of duties following the functions list of the guard at the post and ensures that the prescribed functions of the guards are performed according to prescribed procedure;

3. Conducts redundancy patrol to maximize security visibility and by following the prescribed procedure;

4. Coordinates closely with the unit heads of the different security and business units operating side by side with his detachment;

5. Accomplishes and maintains the Commander's Journal;

6. Accomplishes the Daily Security Operations report for the detachment and submits it to the client representative for discussion, acknowledgement, acceptance and notation before office hours. The Security Operations Report is the summary of all the activities, performances of duties and compliance to all procedure by all guards within the Unit. It is the proof the services of the guards and the agency that employs the guard. A sample of the Security Operations Report is illustrated at Appendix 9.

7. Provides leadership and direction to the security guard force in an emergency following the prescribed procedure:

8. Enforces company regulations;

9. Assists the designated company representative in investigating security incidents;

10. Maintains good rapport with the security unit of the neighboring facilities;

11. Maintains close contact with the concerned heads of government units involved in peace and order maintenance and law enforcement such as the police, military, investigation bureau or homeland security;

12. Conducts continuous security assessment of the facility

and renders report to the client and his superiors for any significant change in the situation;

13. Conducts periodic operational readiness evaluation of his guard;
14. Provides extra instructions to the guard where they are deficient;
15. Performs other security functions as required by the client not covered by routine duties.

DUTIES AND RESPONSIBILITIES OF A TYPICAL SECURITY GUARD

The Guard shall perform the following general duties and responsibilities. They may be distributed to the different post cognizant of the functions or where required:

1. Implement the Security Patrol Procedure

2. Enforce company policies and procedures

3. Implement the Visitor Control Policy Procedure

4. Implement the Control Procedure for Incoming and Outgoing Materials or Properties

5. Maintain custody of all keys entrusted to them and prevent unauthorized handling or duplication

6. Implement the Policy Procedure for Vehicle Control

7. Implement the Emergency Procedures

8. Man the telephones in the absence of the regular telephone operator

9. Report and investigate all unusual incidents to management

10. Maintain a journal of all his operational activities

11. Conduct Duty Turnover Procedure

12. Accomplish and then submit a Daily Security Operations Report to his immediate Officer

13. Perform other functions that his immediate superior may direct him to do

SECURITY UNIT OPERATIONAL READINESS EVALUATION

A team leader must at all times be apprised with the level of readiness of the Security Team. He must have a system to evaluate its readiness. The following are the elements of the Operational Readiness and Performance Evaluation of the Security Team:

1. **Presentable Appearance, Completeness of Uniform and Organizational Equipment**. This element of evaluation is used to determine the completeness, propriety and the demeanor of the guards. These attributes are visual manifestations of the discipline of the security team in abiding by the regulations on uniform as well as how they project the image of respectability and authority.   The uniform also signifies the unity of the team.

2. **Operational Provisions**. This element is used to determine the readiness of the operational equipment of the security unit such as radios, service firearms, metal detectors and watchman's clock are accessible, deployed, operational and well maintained. It also determines the collective proficiency of the team in the use of these pieces of equipment.

3. **Security Operations Records and References**. This element of evaluation is used to determine the discipline of the unit in the proper use, accomplishment and maintenance of records and manuals and to establish their accessibility for reference in the performance of duties, investigation or policy formulation.

4. **Security Reporting and Compliance to Procedures**. This element of evaluation is used to determine the availability and proper use and accomplishment of prescribed reports

and records and to determine the discipline of the security team in collating, filing and timely submission and to determine if the security team performed all the security procedures for the duration of their duties. It also determines if all security incidents, where losses were incurred, have been resolved and were given proper disposition.

5. **Compliance to Contract Provisions**. This element is used to determine if the Detachment had satisfied all the provisions of the service contract and the corresponding Service Level Agreements and that it had complied with all the Laws and Government Regulations. This includes the due remittance of prescribed benefits of the guards and payment of taxes.

A Security Unit Operational Readiness and Performance Evaluation Checklist is illustrated at Appendices 11 to 11c.

SECURITY GUARD OPERATIONAL READINESS EVALUATION CHECKLIST

Aside from the security unit's operational readiness and performance evaluation, the individual guard must also be subjected to evaluation before he assumes duty and periodically for the duration of his tenure. This is to ensure his competency and preparedness to perform and deliver the security requirements of his assigned post. The Guard Operational Readiness and Performance Evaluation used as an instrument for this evaluation is illustrated at Appendix 12 to 12c.

The Guard Operational Readiness Evaluation has the following elements:

1. **Appearance and completeness of uniform and organizational equipment**. This element of evaluation is used to determine the completeness, propriety and the demeanor of the individual guard. These attributes are

manifestations of the guard's self discipline.

2. **Basic security knowledge and operational provisions**.
   This item is used to determine the individual knowledge and
   competency of the guard. A guard has to exercise a certain
   degree of autonomy and manifest reliability. This is the
   reason why they must have a mastery of his function and
   know how to perform them.

3. **Client Evaluation Summary**. This element is used to
   determine the client's descriptive impression of the guard's
   personal conduct which should foster rapport and goodwill
   with the members of the organization.

4. **Attitude and Conduct**. This element of evaluation is used
   to determine the guard's due compliance to all rules of
   conduct and regulations provided for by the Agency as well
   as the Code of Conduct of client organization to which the
   guard is also a subject.

Any negative result on this evaluation at the outset of his
deployment would indicate that the guard is unsuitable. Any negative
result of subsequent evaluation would indicate disaffection and
demoralization. These need to be addressed and corrected, lest the
guard become the security risk.

SUMMARY

Despite the large number of guards all over the world and despite the
criticality of their position and function, guarding appears to be one of the
less developed professions. There is an apparent lack of undertaking to adopt
guarding best practices let alone the adoption of standards.

Guards all over the world are generally regulated by the state. However,
it can be observed that generally, the enforcers are members of either the military
or the police who for practical reasons are more predisposed to enforcement and
reaction to security incidents rather than prevention. The foregoing section of this
book is an attempt to provide the head start for the adoption of best practices for
guarding and its management.

-oOo-

APPENDICES TO SECTION III

Sample forms for recording compliances to security procedures and
security operation activities.

Appendix 1. Security Operations Journal

Security Operations Journal			
Facility: _____ **Page:** _____			
Shift: _____ Time of Duty _____			
Event No.	Time	Observations / Particulars	Remarks
This is to certify that I have read all the entries of the guard on duty and I found them to be in order and I hereby attest to their veracity.			
_____ Guard on Duty		_____ Supervisor	

Appendix 2. Sample Security Patrol Checklist

<table>
<tr><td colspan="10" align="center">Security Patrol Checklist</td></tr>
<tr><td colspan="10">Facility: _____ Page: _____</td></tr>
<tr><td colspan="10">Shift: _____ Time of Duty _____</td></tr>
<tr><td colspan="2" rowspan="2">Control and Critical Points</td><td colspan="6">Patrol Rounds</td><td rowspan="2">Remarks</td></tr>
<tr><td>1</td><td>2</td><td>3</td><td>4</td><td>5</td><td>6</td></tr>
<tr><td>No</td><td></td><td></td><td></td><td></td><td></td><td></td><td></td><td></td></tr>
<tr><td>1</td><td></td><td></td><td></td><td></td><td></td><td></td><td></td><td></td></tr>
<tr><td>2</td><td></td><td></td><td></td><td></td><td></td><td></td><td></td><td></td></tr>
<tr><td>3</td><td></td><td></td><td></td><td></td><td></td><td></td><td></td><td></td></tr>
<tr><td>4</td><td></td><td></td><td></td><td></td><td></td><td></td><td></td><td></td></tr>
<tr><td>5</td><td></td><td></td><td></td><td></td><td></td><td></td><td></td><td></td></tr>
<tr><td>6</td><td></td><td></td><td></td><td></td><td></td><td></td><td></td><td></td></tr>
<tr><td>7</td><td></td><td></td><td></td><td></td><td></td><td></td><td></td><td></td></tr>
<tr><td>8</td><td></td><td></td><td></td><td></td><td></td><td></td><td></td><td></td></tr>
<tr><td>9</td><td></td><td></td><td></td><td></td><td></td><td></td><td></td><td></td></tr>
<tr><td>10</td><td></td><td></td><td></td><td></td><td></td><td></td><td></td><td></td></tr>
<tr><td>11</td><td></td><td></td><td></td><td></td><td></td><td></td><td></td><td></td></tr>
<tr><td>12</td><td></td><td></td><td></td><td></td><td></td><td></td><td></td><td></td></tr>
<tr><td>13</td><td></td><td></td><td></td><td></td><td></td><td></td><td></td><td></td></tr>
<tr><td>14</td><td></td><td></td><td></td><td></td><td></td><td></td><td></td><td></td></tr>
<tr><td>15</td><td></td><td></td><td></td><td></td><td></td><td></td><td></td><td></td></tr>
<tr><td>16</td><td></td><td></td><td></td><td></td><td></td><td></td><td></td><td></td></tr>
<tr><td>17</td><td></td><td></td><td></td><td></td><td></td><td></td><td></td><td></td></tr>
<tr><td>18</td><td></td><td></td><td></td><td></td><td></td><td></td><td></td><td></td></tr>
<tr><td>19</td><td></td><td></td><td></td><td></td><td></td><td></td><td></td><td></td></tr>
<tr><td>20</td><td></td><td></td><td></td><td></td><td></td><td></td><td></td><td></td></tr>
<tr><td colspan="10">This is to certify that I have read all the entries of the guard on duty and I found them to be in order and I hereby attest to their veracity.</td></tr>
<tr><td colspan="5" align="center">_____
Guard on Duty</td><td colspan="5" align="center">_____
Supervisor</td></tr>
</table>

Appendix 3. Sample Visitor's Logbook

Visitor's Logbook

Facility: _____   Address: _____   Page No. _____

Date: _____   Shift: _____

Event No.	Date & Time		Complete Name	Complete Address or Company	Destination	Person to Visit	Purpose	Signature	Remark
	In	Out							

Certified correct: Endorsed by: Noted by:

_____   _____   _____
Supervisor   Security Supervisor   Client Representative

Appendix 4. Sample Incoming Material Logbook

Incoming Material Logbook

Facility: _____   Address: _____   Page No. _____

Date: _____   Shift: _____

Event No.	Date & Time In	Quantity & Unit	Particulars	Supplier Information				Received by	Remarks
				Company	Address/ Telephone	Delivery Receipt No.	Carrier		

Certified correct: Endorsed by: Noted by:

_____   _____   _____
Guard on Duty   Security Supervisor   Client Representative

Appendix 5. Sample Outgoing Material Logbook

Outgoing Material Logbook

Facility: _____   Address: _____   Page No. _____

Date: _____   Shift: _____

Event No.	Date & Time Out	Quantity & Unit	Particulars	Destination Information				Released by	Remarks
				Company	Address/ Telephone	Delivery Receipt No.	Carrier		

Certified correct:

_____   _____
Guard on Duty   Security Supervisor

Endorsed by: Noted by:

_____   _____
Guard on Duty   Security Supervisor   Client Representative

Appendix 6. Unusual Incident Report Checklist

Unusual Incident Report Checklist

Facility _____ Unit:_____

Date Made:	Date Filed:
Name of Originator:	
Name of Recipient:	

Names of other Concerned Recipients	

Classification of Situation or Event: (Check box)	
A Unusual Situation	B. Unusual Event
Safety Hazard	Sickness
Health Hazard	Injury
Fire Hazard	Accident
Unsecured Classified Info	Violation of Regulation
Unsecured Critical Spaces	Violation of Peace and Order
C. Unusual Report	Criminal Incident
Vandalism	
Loss of Property	Time Observed:
Damage to Property	Time Reported:

Place of Occurrence:	

Persons Involved:	A. Perpetrator/Suspect:
	B. Victim/s:
	C. Witness/es
	D. Responder/s:

Estimated Amount of Loss or Damage:

Brief description of the details of the event or situation. (Sequence of events and possible cause.) Use a separate paper.

Action taken in reaction to the event:

Disposition	A. Is the incident resolved?
	B. What is the current status of the case?
	C. Who is now in charge of the case?
	D. Give detail of disposition

_____	_____
Name and Signature of Originator	Name and Signature of Recipient

Appendix 7. Sample Format of an Unusual Incident or Observation Narrative Format Report

---

Unusual Observation Report

Date:
Facility:
Unit:
Address:

To:
Through:
Subject:

1. 1st line Paragraph 1. State here the situation or condition the originator was in when he observed the incident or when he was made aware of the incident.

2. 2nd line Paragraph 1. State here the unusual situation. All elements such as the answers to what, where, when, who, why and how must be stated.

3. Paragraph 2. State the appropriate and necessary courses of action the originator has taken in relation to the observed situation or event.

4. Paragraph 3. The originator shall state here his analysis of the situation or event as to their cause, as applicable.

5. Paragraph 4. The originator shall state here his course of action taken as to the recommend disposition of the case.

_____
Name and Signature of the Originator

Appendix 8. Sample Bomb Threat Checklist Form

Bomb Threat Checklist Form	
Ask:	
What time is the bomb set to explode?	
Where is the location of the bomb?	
How does the bomb look like?	
What explosive materials were used in making the bomb?	
Where was the bomb place?	
How is the bomb activated?	
Record:	
Date and Time of the telephone threat	
Telephone number where the threat was received	
Exact words used by the caller	

Male		Female		Adult		Child		Approximate Age	

Speech:

Slow		Excited		Disguised		Accented		Rapid	
Loud		Broken		Slow		Educated		Normal	
Sincere		Foreign		Uneducated		Signs of Alteration		

Other distinct sounds:

Background Sound

Quiet		Music		Market		Office		Seashore	
Ships		Aircraft		Train		Children		Household	
Traffic		Animals		Machine		Voices		Laughter	

Other distinct sounds:
Name of person who received the call:
Name of person who notified of the call:
Upon receipt of a BOMB THREAT call Security/Authorities Immediately.

Appendix 9. Sample Security Operations Report

Security Operations Report				
Facility: _____ Address: _____				
To: _____ Date: _____				
Guard Service Requirement and Compliance				
	Services & Materials	Required	Actual	Excess (Shortage)
1	Total Services In Hours			
2	Manpower@_____Shifts			
3	Total Absences			
4	Guard Infractions			
5	Agency Officer Visit			
6	Service Fire Arms/Ammo			
7	Others			

Physical Security Inspection, Status and Condition (As Applicable)

1	Surrounding Areas		13	Fire Hazards	
2	Perimeter Fence		14	Safety Hazards	
3	Gates and Doors		15	Damages	
4	Windows		16	Leakages	
5	Security Lighting		17	Material Waste	
6	Building Defects		18	Equipment	
7	Doors to Storage		19	Appliances	
8	Doors to Rooms		20	Classified Info	
9	Fire Fighting Equipment		21	Restricted Areas	
10	Telephones		22	Personal Valuables	
11	Emergency Numbers		23	Security Equipment	
12	Clocks (Synchronized)		24	All Security Posts	

Security Operations

	Events	No.	Reference	Remark
1	Total Visitors		Visitor's Logbook	
2	Incoming Delivery Events		Incoming Material Log	
3	Outgoing Delivery Events		Outgoing Material Log	
4	Security Incidents		Security Journal	
5	Employee Infractions		Delinquency Reports	
6	Vehicle Traffic		Vehicle Logbook	
7	Instructions by Superiors		Security Journal	

Remarks:

Prepared by:

_____
Security Guard Supervisor

Noted by:

_____
Client Representative

Appendix 10. Checklist to Determine a Suitable Agency

	Checklist to Determine a Suitable Agency	Yes	No	Remarks
	Criteria			
1	Does the agency have a valid and duly verified License to Operate?			
2	Has the Agency operated for at least five years with reputable Clients?			
3	Does the Agency have at least 300 guards?			
4	Does the Agency have the same office address as declared?			
5	Does the Agency have the same telephone numbers as declared?			
6	Does the Agency have the present clients as declared?			
7	Does the Agency regularly remit the contribution of benefits mandated by law?			
8	Does the Agency pay the minimum wage to guards as prescribed by law?			
9	Does the Agency pay the required taxes?			
10	Does the Agency have a professional and presentable office?			
11	Does the Agency have a provision for the Retirement Benefit of its guards?			
12	Does the Agency do Drug, Psychiatric, Medical and Physical Tests of its guards?			
13	Does the Agency have an updated 201 file of its guards?			
14	Does the Agency conduct a thorough background investigation of its guards?			
15	Does the Agency conduct Operational Readiness Evaluation for its guards?			
16	Does the Agency conduct periodic re-training of its guard?			
17	Does the Agency conduct daily and nightly inspection of its posted guards?			
18	Does the Agency or its guards submit a Daily Operations Report to their clients?			
19	Does the Agency provide for standard formats for Operations documentations?			
20	Does the Agency Management call on its clients at least every two weeks?			
21	Is the Agency capable of constructing a Guard Force Management Plan?			
22	Does the Agency pay its guards on time?			
23	Does the Agency respond within the day if called upon by its clients?			

Survey conducted by:

Date of Survey:

Appendix 11. Security Unit of Operational Readiness and Performance Evaluation Checklist Part 1

Date: _____ Evaluation Period: _____

Unit:_____ No. of Posts: _____ No. of Guards: _____

Head Guard: _____

Service Requirement in Hours per Day: _____

Type of Facility: _____

Rating for this Evaluation:

	Item	Number of Non-compliance	Base Rating	Score	Weight	Item Rating
A	Completeness of authorized uniform and general appearance		100		.25	
B	Completeness of operational equipment		100		.25	
C	Completeness and proper use of prescribed security records and references		100		.25	
D	Completeness, proper accomplishment and timeliness of submission of required reports in compliance to procedures		100		.25	
Total Rating this Evaluation:						

Appendix 11a. Security Unit of Operational Readiness and Performance Evaluation Checklist Part 2

Appearance and Completeness of Uniform and Organizational Equipment (Place the number of observed non-compliance under the non-compliances column.)				
Presentable/Operational/Applicable				
	Particulars	**No. of Non-Compliances**		**Remarks**
		Yes	**No**	
1	Wearing Head Gear			
2	Has Agency Cap Insignia			
3	Authorized Hair Cut			
4	Properly Shaved			
5	Authorized Uniform			
6	Authorized Name Cloth			
7	Authorized Agency Patch			
8	Authorized Badge			
9	Wearing Agency ID			
10	Buttoned Shirt			
11	Wearing White Undershirt			
12	Wearing Agency Buckle			
13	Wearing Standard Belt			
14	Wearing Standard Whistle			
15	Wearing Standard Holster			
16	Wearing Standard Nightstick			
17	Wearing Black Socks			
18	Wearing Standard Black Shoes			
19	Prescribed Height			
20	Prescribed Weight			
21	Upright Bearing/Countenance			
Operational Provisions: Well Maintained/Operational/Accessible				
1	Valid License			
2	Valid Duty Detail Order			
3	Service Firearm			
4	Basic Ammo Load and Reserve			
5	Appropriate Flashlight			
6	Rain Gear and Boots			
7	Operations Logbook/Journal			
8	Guard Force Management Plan			
9	Metal Detector			
10	Radio Communication			

Appendix 11b. Security Unit of Operational Readiness and Performance Evaluation Checklist Part 3

	Security Operations Records and References: Accessible/Updated		
	Particulars	**No. of Non-Compliances**	**Remarks**
1	Copy of 11 Generals Orders		
2	Copy of Code of Ethics & Conduct		
3	Copy of Client Security Policies		
4	Copy of Specific Functions		
5	Copy of Client Instructions		
6	Copy of Client Procedures		
7	Copy of Fire Procedures		
8	Copy of Bomb Threat Procedures		
9	Use of Prescribed Patrol Checklist		
10	Use of Prescribed Visitor's Log		
11	Use of Prescribed Outgoing Material Log		
12	Use of Prescribed Incoming Material Log		
13	Use of Prescribed Correspondence Log		
14	List of Company Officers		
15	List of Authorized Signatories		
16	List of Emergency Numbers		
17	201 of Guards		
	Security Report and Compliance to Procedures		
	Particulars	**NCO**	**Remarks**
1	Daily Rank Inspection of Guards		
2	Daily Briefing of Guards		
3	Daily Breach Test of Guards or Systems		
4	Submission of Daily Operations Report		
5	Submission of Weekly Operations Report		
6	Submission of Semi-Monthly Operations Report		
7	Submission of Daily Time Records		
8	Submission of Incident Report		
9	Submission of Incident Investigation Report		
10	Regular/Routine Visitation of Agency Officers		

NCO: Non-Compliance

_____
Name and Signature
Security Agency Representative

_____
Name and Signature
Unit Head

_____
Name and Signature
Evaluator

**Appendix 11c. Security Unit of Operational Readiness and
Performance Evaluation Checklist Part 4**

Instructions:

1. Acceptable sampling of guard is at least 20% of the unit strength.
2. Every deficiency noted is scored as 1 NCO (non-compliance)
3. Write NA under NCO for items not applicable.
4. Get the score by subtracting the number of NCO from 100 (base score).
5. Get the item rating by multiplying the score by the weight.
6. Total rating will be the sum of all item ratings.
7. In the spirit of the security principle, "security is good only as its weakest) link", any rating below 100 is deemed not operationally ready and all NCOs must be corrected on the spot or as soon as possible.
8. One copy shall be given each to the evaluator, detachment, facility head and agency.
9. Additional comments and recommendations may be written at the back of the form or in an additional sheet.

Appendix 12. Security Guard Operational Readiness and Performance Evaluation Checklist Part 1

Security Guard Operational Readiness and Performance Evaluation Checklist

Date: _____ Evaluation Period: _____

Name: _____

Place of Duty: _____ Time of Duty: _____

Date of Employment with Agency: _____

Date of Employment at Clients: _____

Evaluation Rating and Classification					
Present Employment Classification		Latest Proficiency Classification (For Regular Guards Only)			
A. Applicant/Trainee		A. Guard 1			
B. Probationary		B. Guard 2			
C. Regular		C. Guard 3			
Item	Raw Score	Maximum Attainable Score	Sub-Rating	Rating Factor	Item Rating
A. Appearance				.15	
B. Operational Readiness				.35	
C. Client/OIC Evaluation				.25	
D. Attitude & Conduct				.25	
Over-all Rating this Evaluation					
A. (Below 50) – Termination					
B. (50-69) – Return to Unit (Retraining)					
C. (70-79) – Probation (on-site Retraining)		*Name of Evaluator*			
D. (80-89) – Regular					
E. (90-99) – Officer Material					
		Name of Rated Guard			

Instructions to Evaluator:

1. Check appropriate space to record observation on the spot.
2. Defects that can be corrected on the spot may be corrected.
3. This checklist may be used for pre-employment evaluation.
4. To set score, count items marked "yes" on both columns.
5. Maximum Attainable Score is the total number of items to be considered in the evaluation. It is the total number of applicable items x 2.
6. Sub-rating is the Raw Score divided by the Maximum Attainable Score.
7. Item Rating is the product of the Sub-rating and the Rating Factor.
8. Overall Rating for the Evaluation is the sum of all the Item Ratings.

Appendix 12a. Security Guard Operational Readiness and Performance Evaluation Checklist Part 2

Name of Guard:				
A. Appearance and Completeness of Uniform and Organizational Equipment				
	Applicable		Presentable/ Operational	
	Yes	No	Yes	No
1. Wearing Head Gear				
2. Has Agency Cap Insignia				
3. Authorized Hair Cut				
4. Properly Shaved Temple				
5. Properly Shaved				
6. Authorized Uniform				
7. Authorized Name Cloth				
8. Authorized Agency Patch				
9. Authorized Badge				
10. Wearing Agency ID Card				
11. Buttoned Shirt				
12. Wearing White Undershirt				
13. Wearing Agency Buckle				
14. Wearing Standard Belt				
15. Wearing Standard Whistle				
16. Wearing Standard Holster				
17. Wearing Standard Nightstick				
18. Wearing Black Socks				
19. Wearing Standard Black Shoes				
20. Prescribed Height				
21 Proportional Weight				
22. Upright Countenance				

Other Defects / Deficiencies / Comments:

Note: Evaluator may deduct one (1) to five (5) points for other defects noted that are not in the list.

Appendix 12b. Security Guard Operational Readiness and Performance Evaluation Checklist Part 3

		Applicable		Well Maintained	
Name of Guard:					
B. Basic Security Knowledge and Operational Provisions					
		Yes	No	Yes	No
1	Valid License and Duty Order				
2	Service Firearm with Ammo				
3	Standard Flashlight				
4	Rain Coat and Boots				
5	Operations Journal				
6	Security Guard Handbook				
7	Metal Detector				
8	Radio Communication				
9	Copy of the 11 General Orders				
10	Copy-Code of Conduct/Ethics				
11	Copy of Client Security Policies				
12	Copy of Specific Functions				
13	Copy of Client Instructions				
14	Copy-Accident Procedures				
15	Copy of Fire Procedures				
16	Copy-Bomb Threat Procedures				
17	Knows Agency Policies				
18	Knows Patrolling Procedures				
19	Knows Guard Basic Functions				
20	Knows Turnover Procedures				
21	Knows Criminal Procedures				
22	Knows Recording Procedures				
23	Can Articulate His Ideas				
24	Knows Force Continuum				
25	Knows Self Defense				
26	Good Public Relation				
27	List of Company Officers				
28	List of Authorized Signatories				
29	List of Emergency Numbers				
Other Defects/Deficiencies/Comments:					

Note: Disregard items, which are not applicable and deduct from the total from the Maximum Attainable Score. Maximum Attainable Score is the total number of applicable items multiplied by 2.

Appendix 12c. Security Guard Operational Readiness and Performance Evaluation Checklist Part 4

Name of Guard:		
C.	**Client Evaluation Summary**	
O	Outstanding	95-99 Points
AAS	Above Acceptable Standard	85-94 Points
MAS	Meets Acceptable Standard	70-84 Points
BAS	Below Acceptable Standard	50-69 Points

	Evaluation	**Point Score**
Knowledge of Job		
Adaptability		
Attitude		
Dependability		
Judgment		
Initiative		
Courtesy		
Cheerfulness		
Appearance		
Total		

D.	**Attitude and Conduct**		
		Counts	**Points**
1. Commission of Light Offenses			
1st Offense		x 1	
1st Repetition		x 2	
2nd Repetition		x 3	
3rd Repetition		x 4	
2. Commission of Serious Offense		x 5	
3. Tardiness		x 6	
4. Excused Absences		x 7	
Total (To be deducted from 99 Points)			

Appendix 13. Sample Vehicle Log and Inspection Checklist

Vehicle Log and Inspection Checklist

No.	Date					Description							Slot No.	Pax	Driver	Remarks
	Time In	Time Out	Type	Make	Color	Plate No.	Accessories					Defects				
							Side Mirror		Aerial	Emblem						
							Left	Right								

Prepared by: Endorsed by: Noted by:

_____   _____   _____

Guard on Duty   *Security Supervisor*   *Client Representative*

Appendix 14. Sample Key Control Card

Key Control Card		
1	Key Number	
2	Lock name or location	
3	Date acquired and date installed	
4	Number of Key duplicates upon acquisition	
5	Number of duplicates with respective dates of duplication	1.
		2.
		3.
6	Distribution of duplicates	1.
		2.
		3.
7	Name of manufacturer, duplicator, address and telephone numbers	
8	Relocation of lock if any	
9	Proof of receipt by person in custody of a key	

Appendix 15. Sample Key Borrower's Logbook

Key Borrower's Logbook

No.	Name of Borrower	Date and time borrowed	Date and time returned	Purpose	Authorized by:	Name of Issuer	Declaration that no duplication was made	Remarks

Prepared by: Endorsed by: Noted by:

_____   _____   _____
Key Custodian   Supervisor   Client Representative

SECTION IV

Security Concepts Application

In this section, the application of all the concepts that were discussed in the previous sections shall be taken up. Some of the topics may be repeated or may simply be referred to. In some instances they will further be expounded to establish their relationship or relevance to the topic on hand.

CHAPTER 10

Developing a Corporate Security Program

A security program provides for the detail on how a security plan for an organization should be carried out. A security program aims to provide the members of the organization a clear understanding of security in relation to the objectives of the organization. It is based on the security philosophy that the organization has adopted. Moreover, it provides the members a clear understanding of their individual and collective roles in security and the rationale for such roles. Lastly, it provides them the guidelines by which their roles are performed. All of these are done with the end in view of developing and instituting a Corporate Security Culture.

DEVELOPING A SECURITY PROGRAM

The following are the steps in developing a Security Program

1. **Adopt a Security Policy**. This policy shall be based on the basic security needs of a team, as stated in Chapter 1. It shall define how the organization plans to address its security needs.

2. **Create and Organize a Security Committee**. The Security Committee shall be the collegial body that will oversee the development and implementation of the organizational security program and the security policies and procedures. The committee shall be composed of the Senior Officers and Unit Leaders of the organization.

3. **Create a Business Integrity Office** that will monitor the execution of the Security Program. This office's main objective is to conduct oversight monitoring of the different

operating units to check for proper execution of the Security Programs of the organization covering Risk Management, Loss Monitoring and Control, Brand or Reputations Security, Environment Security, Operations Audit, Physical Security, Safety, Investigation and the Organizational Security Culture Development.

4. **Designate a full time Business Integrity Officer (BIO)**. This Officer shall head the Business Integrity Office. The Business Integrity Officer must be at an executive level. The Officer shall be accorded the rank, seniority and scope of authority of a senior officer. The Business Integrity Officer shall report to the CEO. The rank and seniority is essential because this will give the Business Integrity Officer the status of being equal with the other members of the management team who formulate policies. As such, the position will make it easy for the Business Integrity Officer to coordinate with the leaders of the other units on the security requirement of their respective teams.

    The authority of this officer should come from the Chief Executive Officer who must provide an explicit policy of such authority.

    The qualification and competencies of the Business Integrity Officer should include a broad spectrum of disciplines such as administration, finance, human resources, operations, communications, public relations, policy formulation, engineering, economics, operations research, project management, education and training, law, investigation on top of his mastery of the fundamental of security and security operation, which cover both proprietary and contractual security units. This officer must possess all the qualities of a good leader and must practice by heart its principles.

The Business Integrity Officer must have a wide network at the command and the operating level of various law enforcement and security agencies of the government.

The Business Integrity Officer must demonstrate having a mind-set of an entrepreneur, forward looking, and who sees security as a means for the company to achieve its goals by prevention of loss rather than by reaction and the preponderance of restrictive and expensive physical security measures.

5. **Write and adopt the Organization's Security Philosophy and the Model by which the philosophy shall be practiced.** This philosophy should embody the belief of the leaders of the organization. It is based on the principles of security upon which all the security programs, policies, procedures and methods should be based. The Security Model shall provide for the aspects of the organization that would be covered by Security.

6. **Designate the team and unit leaders to be responsible for the security program implementation to protect the resources of their respective teams.** Performance of this function must be included in the job description of the leader. Knowledge of security must be a required competency for all leaders.

7. **Cascade the Security Philosophy to all the team members and explicitly state that security of company resources is among their basic responsibilities.** All new hires must have a learning session on security. Involve the team members in all the succeeding steps in the development of a security plan from the determination of the risks and their analysis to their suggestions on how to mitigate those risks. The learning should cover the Basic Security Concept, the Security Philosophy of the Organization, the Basic Objectives of the Organization, the eleven classifications of resources, and the threats and risks that

confront the organization and its security needs.

8. **Conduct a Security Assessment**. This step involves Asset Criticality Assessment, Risk Assessment, Vulnerability Assessment and Acceptability. This should be a corporate wide undertaking where every member of the team is involved. The processes shall be explained in detail in this chapter.

9. **Create a General Security Plan for the Organization**. Create specific security plans, policies and procedures to address specific security risks or condition. It is also important to involve the team leaders and members in the formulation of the security plans. The involvement of team members will foster their sense of ownership of the plan and their involvement with their implementation.

10. **Determine the Security Awareness Level of all the Employees.** Survey and monitor the members' awareness and practice of security. The recommended awareness index of all team members in an organization should be at least 90%. An individual security awareness level of A means the team members have internalized Security and it is manifested through practice. It also means that the members can actively contribute to the conceptualization of security measures. Security Awareness Survey for the team members is a tool for determining the security awareness level of the individual member as well as the organization. This tool can be formulated in-house although there are available tools presently being used by security consultants.

11. **Write a Security Manual for Procedures**. This manual is for general applications such as security incident reporting procedure and investigation procedures. Security procedures that are peculiar to the unit or user may be imbedded in the

operating policies. Together with the creation of the manual is the creation of the different forms to document the compliance to the established security processes. The forms or templates must cover all the information gathered from the performance of a certain security function or procedure. Appendices 1-12 in Section 3 illustrate examples of these forms.

12. **Monitor the cost of losses of the company caused by untoward (security) incidents, mishaps, missed opportunities, penalties and wastage.** This should be compared with previous periods when there was no established security program. Determining the actual losses of the company and monitoring it will provide the basis to determine the effectiveness and cost efficiency of the security program.

Summary

A successful security program starts with the involvement of the leadership of the organization. The leadership must set the policy for its adoption. All team members must be enjoined in the assessment of the risks and creation and enforcement of security of policies and procedures.

-oOo-

CHAPTER 11

Security Assessment and Security Plan

The first step in the development of a **Security Plan** is to make a **Security Assessment**. Security Study and Assessment is a series of processes to determine the security condition of a facility or more importantly, the organization. It describes the situation upon which the security plan is made. This is similar to a related discipline called Risk Management.

The fundamental difference between Security Assessment and Risk Management is that the latter covers a much wider scope in defining risks and the factors for their probability to occur. It covers hazards, threats or risks which are abstract and whose impact cannot be quantified. Examples of these risks that a security assessment covers are absence of security culture in an organization, employee ignorance, low morale and employee indifference.

Since one of the purposes of this book is to integrate all disciplines whose objective is to preserve and protect the resources of the organization with the end in view of making the organization achieve its objectives, the term Security Assessment shall be used. This is because Security Assessment has a much broader scope and it utilizes additional tools in making detailed assessment.

A Security Assessment involves the following activities:

1. **Asset Criticality Assessment**. This is a process, which involves the identification, assessment and rating of all the company resources as to their relevance, criticality or their impact to the organization should they be destroyed, lost or rendered inutile.

2. **Hazard and Threat Definition**. This is a process where all the hazards and threats to which the organization and its resources are susceptible to are determined. This process also involves the determination of how critical the identified hazard is.

3. **Mishap Occurrence Likelihood Analysis**. This is a process of determining and rating the likelihood of a mishap to occur based on the obtaining factors and conditions surrounding the asset.

4. **Asset Vulnerability Assessment**. This is an activity, which involves the determination and rating of how vulnerable the resources of the organization all to the specific hazards that threaten the identified resource.

5. **Risk Assessment**. This is the process of determining and rating the degree of risk that threatens an organization and its resources.

6. **Risk Acceptance**. This is the process of deciding which of the assessed risk and conditions are acceptable or not to the organization. This process also involves the making of priorities as to which of the rejected risks shall be avoided, eliminated or mitigated.

The above set of processes is similar to the common organizational jargon called SWOT (Strengths, Weaknesses, Opportunities and Threats) Analysis, which requires an organization to know its capabilities and that of its competitor, whether it is physical or abstract. These processes are also described in detail in the following topics. The security assessment shall provide for the Situation in the General Security Plan. Security Assessment implies one basic principle of warfare, that is, "Know yourself and know your enemy."

Before going to the details on how to conduct a Security Assessment, it is essential to know the terminologies that are essential in understanding the entire process. Several of these terms have been defined in previous chapters but they are herein reiterated for recall. There are other terminologies that will be introduced in the description of the different processes of the security assessment. Among these terms are as follows:

1. **Cost**. Cost is the monetary price for which an asset or resources is acquired. This is generally used in Risk Management.

2. **Value**. Value refers to the essence of a particular resource to the organization. Often, this is commonly equated with the cost as defined previously. However, there are assets that cost less than the other but nevertheless have equal, if not more value than the other in terms of function, utility or productivity. For example, the cost of a telephone line cost is lot less than a vehicle but, to a trader their values are the same; the absence of one or the other will be detrimental to the business in equal terms.

3. **Relative Value**. Relative Value refers to the degree of value given to a resource or asset of the same kind and price by different entities. The telephone and the vehicle can illustrate this also. While the difference in cost of each from the other is great, their relative values may be the same. It can also be illustrated in another way by a fixed amount of investment, say US$ 20,000. The loss of that amount may be insignificant to a car manufacturer but the same amount of loss may mean the closure of a "mom and pop" store.

ASSET CRITICALITY ASSESSMENT

The fist step in a Security Assessment is a process called Asset Criticality Assessment. Criticality Assessment is the process of determining the degree of impact to the organization when a resource is destroyed or lost as a result of a hazardous incident or mishap.

Criticality Assessment can be applied to assets or to hazardous events. When applied to assets, it is called an Asset Criticality Assessment. The practitioners of conventional Risk Management call this process Business Impact Analysis.

As discussed in Chapter 1, the resources of the organization are the means by which the objectives are achieved. As such, they are the reason why security is needed in the first place. As a basic guideline, anything that a company spends for is a resource. To reiterate these resources, they belong to any of the following classifications:

1. Financial Resources
2. Human Resources
3. Facility
4. Equipment
5. Materials
6. Product or Services
7. Market
8. Environment
9. Information
10. Reputation
11. Time

HAZARD DEFINITION AND CRITICALITY ASSESSMENT

Criticality Assessment can also be applied to the security hazards and risks that threaten the resources of the organization. Security hazards or risks have different degrees of impact on assets. Some security hazards or risks are more devastating than others. Some security hazards can affect more resources than the others, while some security hazards and risks can only threaten a specific resource.

To reiterate these hazards, they belong to either, Natural Hazards or Man-made Hazards the examples of which are listed in Chapter 1.

In defining the hazards and risks that could threaten the team's resources, it is essential to be specific rather than use a general term. This is because every specific hazard has different impacts or they could cause different types of damage as to their extent or as to the susceptibility of resources to them. Example: General Term – Natural calamities, Specific – typhoon, earthquake, etc.

There are three categories of resources for which Criticality is conventionally applied. They are Human, Reputation and Finance. The impact or severity of their loss is ranked according to the cost, extent of the damage or duration and extent of publicity the mishap elicits.

Usually, ranking is graduated into five ranks from highest to lowest. Each rank has a corresponding numerical weight, which will be used as a factor in determining the magnitude of the risks.

For material assets, the severity is measured by adding all the relative and incidental costs to the present cost of the asset lost. The term Criticality can be used for this purpose. Criticality is determined by the following formula:

$$X = (X_r + X_s + X_o + X_i) - (I - P)$$
Where:

X - Criticality
X_r - Cost of replacement
X_s - Cost of temporary substitute
X_o - Other Related cost
X_i - Cost of income loss
I - Indemnity by insurance
P - Amount of Insurance Premium

Other related costs can be any of the following attributes necessary for a more accurate means of describing the criticality of an asset.

1. Mission impact
2. Customer impact

3. Safety and environmental impact
4. Ability to isolate single-point-failures
5. Preventive maintenance history
6. Corrective maintenance history
7. Average time between failures
8. Likelihood of failure
9. Spares lead time
10. Asset replacement value
11. Planned utilization rate

The conventional manner by which the impact of loss is ranked is that the next lower rank is 10% of the previous estimated cost. The highest cost, whose loss is deemed extreme, shall be used as the basis. This depends on how the organization would put the value. The value could be the value of the entire business itself, the value of the capital investments or 50% of its net worth. So, for Financial Impact where the deemed cost of extreme ranking is $ 100 Million. The ranking are in the following table together with the ranking for Human and Reputation Impact.

Category	Extreme (5)	Very High (4)	High (3)	Moderate (2)	Low (1)
Financial	$100 M or over	$50M-99 M	$1 M - 49M	$100K - 999K	$99k or below
Human	Multiple Casualties	Single Casualty	Multiple Injuries	Serious Injury	Minor Injury
Reputation	Worldwide Publicity	Regional Publicity	National Publicity Extended	National Publicity Instance	Industry Publicity Instance

Figure 11. Ranking for Financial, Human and Reputation Impact. *The variances between ranks are subject to the decision of the organization. The common practice for financial impact is that the next lower rank is 10% of the higher rank; while the variances between Human and Reputation are based on best practices*

The Ranking of Criticality of Assets may be assigned a description as follows:

1. **Impact Unknown.** This is assigned due to insufficient data. This rating is given as an interim description until sufficient data is acquired to give a permanent rating.

2. **Low**. Low Impact Rating means that the impact of loss is relatively unimportant and it can be covered by contingency reserves.

3. **Moderate**. Moderate Impact Rating means that the impact of loss is moderately serious and it would result to a negative impact on the earnings.

4. **High**. High Impact Rating means that the impact of loss is serious and it would result to a noticeable impact on the earnings.

5. **Very High**. Very High Impact Rating means that the impact of loss is very serious and that a certain loss or damage would require a major change in investment policies.

6. **Extreme**. Extreme Impact Rating means that the loss is fatal to the business and that there is a need for a total re-capitalization or abandonment of the business.

Those assets identified to have serious and higher impact would thus be the assets or risks that should be given priority for recognition by management for further assessment for vulnerability and the likelihood of the mishaps that threaten them. Any security measure to be used should consider the inherent vulnerability against a security risk with high likelihood of occurring to minimize the cost of security.

If all the team members of all the units in the organization were to be involved in the security assessment, they should be educated in the fundamentals of security rather than be given a template to accomplish. The team members of subordinate units should limit their assessment to the resources of their own units. It should be understood that the severity or criticality rating of the unit's resources might not be the same as the criticality of the same resource in relation to the entire organization. This condition is what is called **Relative Criticality**. Relative Criticality means that the criticality of same resources or hazard varies from one unit to another or even to the organization itself.

MISHAP OCCURRENCE LIKELIHOOD ANALYSIS

Mishap Occurrence Likelihood Analysis is the process of determining the likelihood of a particular mishap to occur.

This process includes the determination and consideration of other internal or environmental factors that can cause a mishap to occur. Here lies another fundamental difference between Security Assessment and Risk Management. The basic, and often, the only factor considered by a Risk Management Practitioner is history. This means that the likelihood for a hazard to occur is determined by the frequency of its past occurrences. This is because history is the only factor that is certain and the only factor that can be quantified. To a Security Practitioner as applied in the context established in this book, there are a number of factors in determining the likelihood for a hazard to occur.

The following table identifies those factors, their source and their attributes.

	Factor	(+) Favorable Factor	(-) Unfavorable Factor
A	Knowledge, Competence	Training and Education	Ignorance or Ineptitude
B	State of Morale and Discipline	High Morale, Discipline	Demoralization, Disaffection
C	Values	Positive Values	Negative Values
D	Mental and Emotional Health	Mental and Emotional Stability	Mental Derangement
E	Physical Configuration	Bulky, Heavy, Immovable	Small, Light, Portable
F	Street Value	No Takers	Salable to Fencers
G	Security Policies	Relevant Security Policies	Absence of Security Policies
H	Doctrines and Principles	Synergy of Doctrines and Principles	Lack of Synergy of Doctrines and Principles
I	Staff Function	Synergy of Staff Function	Lack of Synergy of Staff Function
J	Security Measures	Presence of Security Measures	Absence or Inadequacy of Security Means
K	Protective Structures	Adequate Protective Structures	None or Defective Protective Structures
L	Emergency Equipment	Presence of Emergency Equipment	None or Defective Emergency Equipment
M	Proximity of Law Enforcers	Near a Police or Fire Station	Far from Police or Fire Station
N	Social and Economic Condition	Favorable Socio-Economic Condition	Unfavorable Socio-Economic Condition
O	Political, Ideological	Absence of Insurgency or Rebellion	Presence of Insurgency or Rebellion
P	Historical Events	No Past Event	Recorded Past Event
Q	Criminal Capability	Amateur, Small time, Unsophisticated	Professional, Well Funded, Sophisticated
R	Geographic Location	Favorable Geographical Location	Unfavorable Geographical Location

Figure 12. Factors for Susceptibility and Likelihood of Outcome for Operational Risks
The factors for Susceptibility and Likelihood of Occurrence are not limited to the above list.

Their presence or absence of the stated factors and their combination can influence the likelihood of occurrence or non-occurrence of a security incident or a mishap.

A model for **Operational Risk Occurrence Likelihood Analysis** as illustrated at Figure 14 illustrates a simple way to determine the likelihood for a Mishap to occur. The total number of positive factors is compared with the total number of negative factors. The more positive factors there are than negative factors, the lower the likelihood of the mishap. Conversely, the more negative factors there are than positive factors, the higher the likelihood for a mishap to occur. There are conditions however, that a single factor should be enough to make a mishap to occur with certainty. An example of this is typhoon or earthquake for those facilities that are located in the typhoon belt or the Pacific Ring of Fire, respectively.

The Mishap Occurrence Likelihood Rating shall then be assigned to each defined risk or hazard together with the assets to be affected by the same Security Risk should they occur. The following are the Likelihood Ratings and their respective descriptions.

Rating	Likelihood	Description
4	Certain	The mishap will occur
3	Likely	The likelihood of mishap is much greater than its non-occurrence
2	Moderately Probable	The mishap is more likely to occur than not to occur
1	Unlikely	The mishap is less likely to occur than to occur
0	Indeterminate	Insufficient data for evaluation

Figure 13. Operational Risk Outcome Likelihood Rating

Determine the Risk for each asset by using the Risk Rating Matrix. Risk here is the potential for damage, loss or harm to people, assets, environment and reputation. It is the combination of threat and their potential impact on the asset and the likelihood of occurrence. Risk, for the purpose of obtaining its rating, is the product of the equivalent numbers designated to provide value to the Criticality (Severity) Rating of the assets and the Security Incident Occurrence Likelihood Rating of the Security Risks. A sample Risk Rating Matrix is illustrated at Figure 15. There are other variations of the Risk Rating Matrix being used by different Risk Analysts.

Rating	Security Event/ Hazard	Asset Affected	Risk Occurrence Factors																		Likelihood
			A	B	C	D	E	F	G	H	I	J	K	L	M	N	O	P	Q	R	
1	Accident	Personnel Morale Facility Equipment Stocks Income	+	+	+	+	NA	NA	+	+	NA	NA	+	+	NA	NA	NA	−	NA	NA	Unlikely (1)
2	Earthquake	Personnel Equipment Facility Income Morale Image	NA	NA	NA	NA	NA	NA	NA	NA	NA	NA	NA	NA	NA	NA	NA	−	NA	−	Certain (4)
3	Employee Dishonesty	Income Morale Image	+	−	−	−	+	−	+	+	NA	+	+	+	NA	−	NA	−	−	NA	Likely (3)
4	Terrorism	Personnel Facility Equipment Information Income Morale Image	NA	NA	NA	NA	NA	NA	+	NA	NA	I	I	+	−	−	−	+	−	+	Indeterminate (0)
			A	B	C	D	E	F	G	H	I	J	K	L	M	N	O	P	Q	R	

Legend (+) Favorable or Provided; (−) Unfavorable or No Provision; (I) Inadequate; (NA) Not Applicable; (U) Unknown (No Observation/No Record)

Figure 14. Model for Security Incident Occurrence Likelihood Analysis. *The more negative factors there are in a given condition, the more likely that a security incident can occur. The alphabet under the Risk Occurrence Factor represents a specific factor as described in the Risk Factor Table at Figure 12.*

Each asset whose criticality or severity has been determined and the security risks whose likelihood to occur have been determined should be superimposed on the Risk Rating Matrix to determine the Risks to the company should it be lost or damaged. .

The determination of the Risk should not be the only basis of the organization in deciding the need to spend resources to mitigate the risk or its impact. It must first be determined which of the risks can be eliminated without cost. For example, absence of security culture and ineptitude can be eliminated by education and training.

Thereafter, identify the remaining risks with high ratings and determine if they can be avoided. For example, typhoon is a risk that cannot be eliminated. However, the organization can avoid the risk by not going to locations that are along natural typhoon paths.

S **E** **V** **E** **R** **I** **T** **Y**	Extreme	5	5-H	10-U	15-U	20-U
	Very High	4	4-M	8-H	12-U	16-U
	High	3	3-L	6-M	9-H	12-U
	Moderate	2	2-L	4-M	6-M	8-H
	Low	1	1-I	2-L	3-L	4-M
U - Ultra			1	2	3	4
H - High			Unlikely	Moderately Probable	Likely	Certain
M - Moderate						
L - Low			**LIKELIHOOD**			
I - Insignificant						

Figure 15. Risk Rating Matrix. Risk Rating is determined by the product of numerical rating of the severity of loss of a particular resources and the likelihood of a specific mishap to occur. Those with the ultra risk require that they must be addressed with top priority.

The next step is to determine the risks that cannot be eliminated nor avoided. Then, the assets or resources that are susceptible to those risks are determined. Asset Susceptibility is a condition where resources can be directly affected by or can be subject of the risks because of their nature. For example, inanimate resources cannot be susceptible to epidemic in the same way that humans cannot be subject to depreciation.

ASSET SUSCEPTIBILITY ANALYSIS

Asset Susceptibility Analysis is the process of determining which assets are likely to be affected by a specific hazard or risk. Some assets are more likely to be affected by some hazards than others. For example, a cellular phone is more likely to be stolen than an air conditioner because the former is portable and "pocket–able." Facilities at the tropical zones are more susceptible to typhoon than those facilities located in the torrid zones of the earth. Assets which are not susceptible to a specific hazards or risk need not have to be considered for security intervention against that hazard. A sample Facility Threat Susceptibility Table is illustrated at Figure 22.

ASSET VULNERABILITY ANALYSIS

Organizational resources that are susceptible to the risks should then be subjected to a process called Vulnerability Analysis

Asset Vulnerability Analysis is the process of determining the presence, absence or effectiveness of security measures applied to the organization and its specific assets. It is usually conducted with the Risk Occurrence Likelihood Analysis. Vulnerability Assessment is not limited to identifying the security measure but it includes the security risk, to which the asset is susceptible and vulnerable to. The security measure or need can be more than one.

The ratings for the Vulnerability Analysis are as follows:

Risk/Hazard	Asset Affected	Measures (4Ps)	Provision	Documented	Vulnerability
Accident	Personnel Morale Facilities Equipment Stocks Income	Personnel Security	Y	Y	Low (1)
		Policies & Procedures	Y	Y	
		Physical Security	Y	Y	
		Practices	Y	Y	
Earthquake	Personnel Facilities Equipment Stocks Environment Income Reputation	Personnel Security	I	I	Indeterminate (0)
		Policies & Procedures	U	I	
		Physical Security	U	U	
		Practices	Y	U	
Employee Dishonesty	Income Morale Reputation	Personnel Security	Y	Y	Low (1)
		Policies & Procedures	Y	Y	
		Physical Security	I	Y	
		Practices	Y	I	
Terrorism	Personnel Facilities Equipment Materials Information Labor Stocks	Personnel Security	N	N	Extreme (4)
		Policies & Procedures	I	N	
		Physical Security	N	N	
		Practices	N	N	

Y–Yes, N–No, I–Inadequate, U–Unknown, NA–Not Applicable

Figure 16. Sample Facility Vulnerability Analysis. The more security measures are present to address the hazards, the less chance that a resource will be affected, therefore the lower the vulnerability.

Code	Vulnerability Rating	Description
4	Extreme	No protection or mitigation at all.
3	High	The means of protection is/are ineffective.
2	Moderate	The means of protection is/are relatively effective.
1	Low	The means of protection is/are considerably effective.
0	Indeterminate	The effectiveness of protection is unknown.

Figure 17. Vulnerability Rating Description Table

SECURITY SURVEY AND SECURITY AUDIT

Security Surveys and **Audits** are the precursors of Security Assessment. These are the means of gathering information from the target facility in regard to its security. There can be no Security Assessment without a Security Survey. There are two basic tools for a Security Study and Assessment. They are the Security Survey and the Security Audit.

A **Security Survey** is an exhaustive examination and thorough inspection of all assets, operational systems and procedure within an installation. It is done when no previous security assessment has been done on a facility or an organization. Its purposes are the following:

1. To determine the existing state of security;
2. To locate the weaknesses in the defense;
3. To determine the degree of protection required;
4. To create solution to identified weaknesses.

A survey is done on the following circumstances:

1. When starting up a business;
2. When putting up a new facility;
3. When transferring to a new facility or location;
4. When there is a major change in organizational structure or business direction.

A **Security Audit** is done to determine the compliance to the recommended measures provided by the Security Plan and their effectiveness so that adjustments can be made. This should be done periodically, the frequency of which should not be less than one year.

JOEL JESUS M. SUPAN

DETERMINING ACCEPTABILITY OF RISKS AND CONDITIONS

The last step in conducting a Security Assessment is the statement of "**acceptability**" or "**unacceptability**" of a condition. This step is not covered by conventional assessments. However, it is deemed necessary, as it will guide the team members on the position the organization has taken in addressing the said risks.

"**Acceptable**" conditions are those conditions that are deemed insignificant or low risks. They can also be the conditions with higher risk but with adequate security measures to mitigate the said risks.

On the other hand "**unacceptable**" conditions are those conditions that are found to be ultra, high or medium risks with inadequate protection.

The "unacceptable" conditions shall provide the bases for preparation of a Security Plan which shall be constituted by the different security measures applicable to a specific risk with due consideration to cost.

Acceptability precedes the tolerance that the organization allows for a particular risk or condition, surrounding it and the present level of protection. It is determined with due consideration to the result of the criticality, risk and vulnerability assessment.

Risk Acceptance is the process in which management would decide on what condition is acceptable to the organization or not. The factors or conditions for acceptability can be any of or a combination of the following:

1. Inevitability of a hazardous event
2. Cost of security that could exceed the benefits of the asset to be protected.

It is upon acceptance of the risk by which the organization decides on what specific security measures shall be applied to mitigate, if not eliminate the risk as assessed.

The following Risk Assessment and Acceptance Table (Figure 18, is a tool that summarizes is simple and understandable statement of a Security Assessment Security Assessment, the acceptability of the obtaining conditions and the recommended actions.

Security or Loss Incident	Asset Affected	Likelihood of Incident Occurrence	Criticality	Vulnerability	Risk	Acceptance	Recommended or Priority Action
Accident	Personnel Morale Facilities Income Equipment Stocks	Low (1)	Low to Moderate (1-2)	Low (1)	Low (2)	Acceptable	Implement, Enhance Emergency Response Capability
Typhoon	Personnel Morale Stocks Facilities Income Equipment	Certain (4)	Moderate to High (2-3)	Undetermined (4)	Ultra (20)	Not Acceptable	Prepare a Typhoon Doctrine
Member Violation	Income Morale Reputation	Likely (3)	Low to Moderate (2)	Low (1)	Moderate (6)	Not Acceptable	Maintain, Enhance Pre-Hiring, Screening Due Diligence Supervision
Terrorism	Personnel Facilities Stocks Equipment Information Labor Materials Reputation	Indeterminate (0)	Extreme (5)	High (4)	Ultra (20)	Not Acceptable	Develop and Implement an Anti-Terrorism Doctrine for an Integrated Security Program

Figure 18. Sample   Risk Assessment and Acceptance Matrix. Note: Acceptance of a condition is a Management Decision

The following flow charts at Figure 19, 20 and 21 illustrate the processes of Risk Assessment, Vulnerability Assessment and the Risk Mitigation Framework.

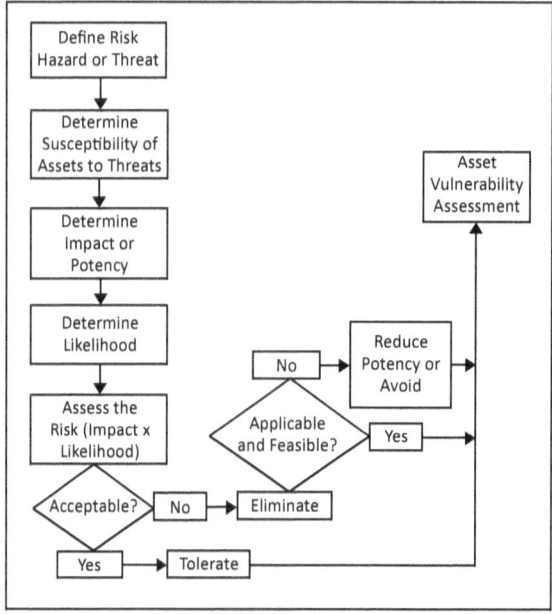

Figure 19. Risk Assessment Flow Chart Diagram

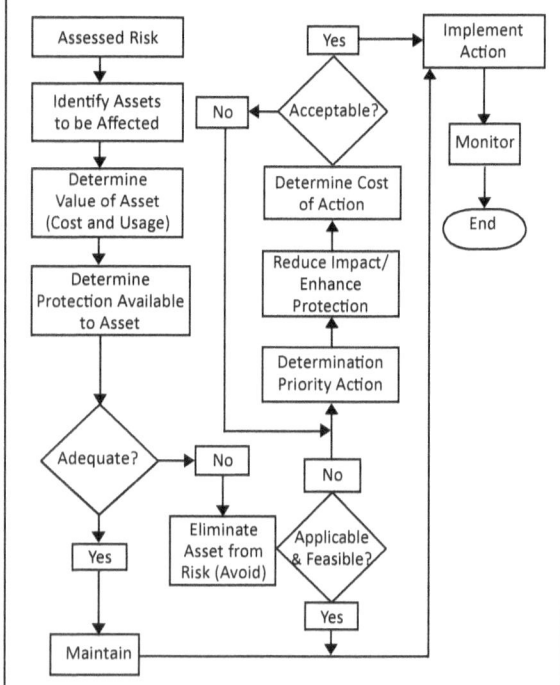

Figure 20. Asset Vulnerability Assessment Flow Chart Diagram

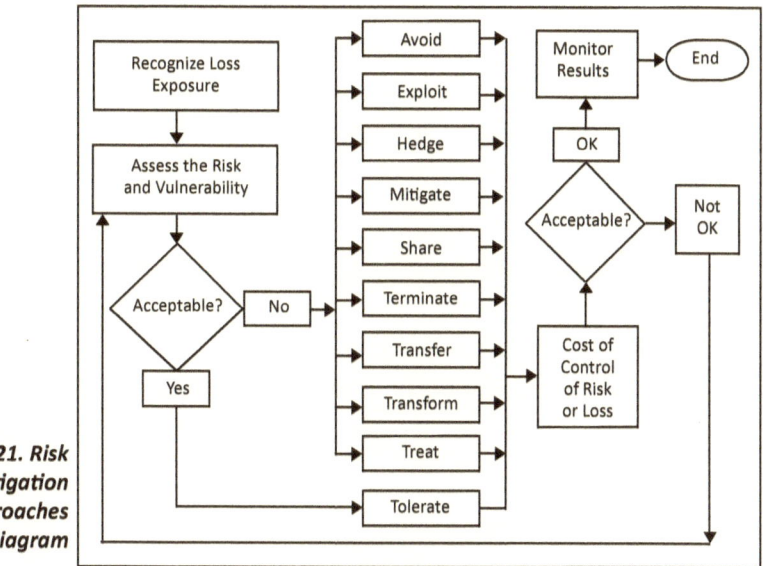

Figure 21. Risk Mitigation Approaches Diagram

DEVELOPING A GENERAL SECURITY PLAN

A **General Security Plan** provides for the overall scheme in securing the resources of the organization. Like all plans, it must be written and it must have provision on the following basic elements:

1. **Situation.** This element provides for the estimates and assessment of the condition surrounding the resources to be protected in the organization. It is basically the result of the Security Assessment. It provides the rationale and justification to develop a security plan. It covers Asset Criticality Assessment or Business Impact Analysis, Risk Assessment, Vulnerability Assessment and Acceptability.

2. **Objective.** This element shall provide the desired result of the security plan relative to the assessed condition.

3. **Implementing Guidelines.** This element of the security plan shall provide the policies and guidelines to be followed in addressing the assessed condition. It shall specify the specific project to be made to address the significant risks that were identified.

JOEL JESUS M. SUPAN

4. **Administrative and Financial Support**. This element provides the required support mechanisms such as manpower, equipage, supplies and materials and finances to implement the General Security Plan. These can be derived from the individual and peculiar security plans and projects created to address specific risks or condition.

5. **Organization for Management and Control**. This element shall provide for the organization, authority, responsibility and tasking of persons involved to carry out the plans. It shall also provide for the lines of communications, the monitoring, control mechanism and reportorial requirements,

The General Security Plan shall have for its appendages the individual security implementing plans which will address the specific security conditions.

The General Security Plan shall be based on the various security assessments to be conducted by the different units of the organization. The tools for these assessments are embodied in various management undertaking such as Business Intelligence, Competitive Intelligence, Risk Management (ERM) and the Business Continuity Plan (BCP), one component of which is Disaster Preparedness, Response and Recovery Plan (DRP).

-oOo-

CHAPTER 12

Security System Design and Application

This chapter will provide the examples and guidelines on how to formulate a security program based on what was taken up in all the preceding chapters.

SECURITY CONCEPTS APPLICATION

Security concepts application simply means putting to use and practice all the concepts covered in this book. The industry is replete of publications on specialized and specific security methods and programs for the different types of environment and industry examples of which are as follows:

1. Airline Security
2. Airport Security
3. Bank Security
4. Building (Office) Security
5. Building (Residential) Security
6. Camp Security
7. Campus Security
8. Cash Transport Security
9. Communications Security
10. Construction Security
11. Corporate Security
12. Critical Infrastructure Security
13. Diplomatic Mission Security
14. Executive Protection
15. Home Security
16. Housing Security
17. Homeland Security

18. Hospital Security
19. Information Security
20. Information Technology Security
21. Mall Security
22. Manufacturing Facility Security
23. Mining Site Security
24. Museum Security
25. Office Security
26. Personal Security
27. Power Plant Security
28. Railway Security
29. Railway Station Security
30. Refinery Security
31. Retail Security
32. Seaport Security
33. Transport Terminal Security
34. Vehicle Security
35. Vessel Security
36. Water Reservoir Security

The differences in the security systems and measures used in these facilities are defined by the hazards or risks that threaten them. In turn, the hazards and risks are defined by the type of critical resource they have that they need to protect first and foremost. The detail of the threats to which the facility is susceptible to is illustrated at Figure 22.

The table shows that the basic differences between all the types of security are the varying degrees of usage of the physical security elements and the corresponding protocols and procedures.

But regardless of the differences of the types of security measures applied to the facilities, the principles and the aspects, needed to protect them from the threats and risks are the same.

For example Bank Security's major threat is robbery, fraud and fire because its most critical assets are its people, the facility and the cash.

Figure 22. Facility Threat Susceptibility Table

Facility	Resources or Assets Peculiar to the Facility	Likely and Peculiar Hazards and Threats to which the Facility is Susceptible
Air Line	Passengers, Crew, Airplane, Baggage	Natural Phenomenon, Terrorism, Hijacking, Weather, Sickness, Theft, Fire, Human Error, Labor Unrest, Aging Personnel, Aging Machine
Airport Terminal	Passengers, Workers, Buildings, Facility, Baggage, Cargo, Runway, Equipment	Natural Phenomenon, Terrorism, Hijacking, Weather, Sickness, Extortion, Theft, Fire, Human Error
Bank Office	Client, Employees, Cash, Equipment, Appliances, Records, Documents	Natural Phenomenon, Fraud, Embezzlement, Robbery, Sickness, Theft, Fire
Bus line and Terminal	Passenger, Crew, Documents, Bus, Environment, Cargo, Baggage, Equipment,	Natural Phenomenon, Ageing Buses Ageing Crew. Extortion, Fire, Hijacking, Human Error, Robbery, Sickness, Theft, Pick pockets, Terrorism
Camp	Personnel, Building, Vehicles, Machinery, Ammo Dams, Armories, Weaponry	Natural Phenomenon, Enemy Attack, Sabotage Infiltrations, Sickness, Theft, Fire
Communications	Tower, Equipment Transmission Lines	Natural Phenomenon, Intrusion, Terrorism, Extortion, Vandalism, Fire, Theft, Fire.
Corporation	Personnel, Information Equipment Facility	Natural Phenomenon, Extortion, Fire, Fraud, Theft, Sickness, Injury, Vandalism, Loss of Information,
Diplomatic Mission	Diplomats, Staff, Citizens, Cash, Equipment, Appliances, Records, Documents	Natural Phenomenon, Robbery, Sickness, Theft, Fire, Espionage, Terrorism
Executive	Personnel, Family, House, Documents, Information, Personal Assets	Assault, Extortion, Kidnapping, Robbery, Sickness, Accident, Injury, Theft
Factory or Manufacturing Facility	Employees, Building, Record, Environment Machinery, Equipment Materials Products	Natural Phenomenon, Sickness, Accidents, Substance Abuse, Natural Disasters, Infestation, Fire, Vandalism, Assault, Kidnapping, Theft, Robbery, Labor Unrest
Home and Housing	Family Members, House, Document Personal Valuables, Appliances, Furnishing	Natural Phenomenon, Assault, Extortion, Kidnapping, Robbery, Sickness, Accident, Injury, Theft, Fire, Infestation

Figure 22. Facility Threat Susceptibility Table

Facility	Resources or Assets Peculiar to the Facility	Likely and Peculiar Hazards and Threats to which the Facility is Susceptible
Homeland	Citizens, Critical Infrastructure, Natural Resources Industry, Commerce, Agriculture	Natural Phenomenon, Terrorism, Natural Disasters, Economic Sabotage, Criminality, Civil Unrest, Invasion, Poverty, Economic Recession
Hospital	Patients, Staff, Records, Equipment, Medical Supplies	Natural Phenomenon, Fire, Theft, Skipping, Vandalism
Information and Document	Storage, Equipment, Records	Sabotage, Theft, Fire, Infiltration, Corruption
Information Technology	Storage, Equipment, Records	Sabotage, Theft, Fire, Infiltration, Hacking, Virus, Obsolescence
Maritime Port Terminal	Passenger Crew, Building, Cargo, Baggage, Environment	Terrorism, Hijacking, Sickness, Extortion, Theft, Fire, Pickpockets
Maritime Vessel	Passenger, Crew, Ship, Records, Environment, Cargo Baggage	Terrorism, Hijacking, Hostage, Taking, Weather, Sickness, Theft, Fire, Human Error, Aging Personnel, Ageing Vessel
Mining Site	Employees, Records, Artworks, Artifacts, Building Equipment	Natural Disasters, Sickness, Accidents, Substance Abuse, Infestation, Fire, Vandalism, Robbery, Assault, Robbery, Kidnapping, Theft, Labor Unrest
Museum	Employees Records, Artworks, Artifacts, Building, Equipment	Natural Phenomenon, Fire, Theft, Skipping, Vandalism
Nuclear Plant	Employees, Building, Records, Machinery, Materials, Equipment Environment	Sickness, Accidents, Assault, Substance Abuse, Natural Disasters, Infestation, Fire, Vandalism, Kidnapping, Theft, Robbery, Labor Unrest, Meltdown
Office Building	Occupants, Building, Equipment, Vehicles	Natural Phenomenon, Fire, Terrorism, Hijacking, Sickness, Theft
Personal (Executive)	Personnel, Family, House, Document, Personal Asset, Appliances	Natural Phenomenon, Assault, Extortion, Kidnapping, Robbery, Sickness, Accident, Injury, Slander, Theft
Power Plants	Employees, Residents, Building, Records, Materials Machinery, Equipment, Environment	Natural Phenomenon, Fire, Sickness, Accidents, Theft, Substance Abuse, Assault, Infestation, Vandalism, Kidnapping, Robbery, Labor Unrest

Figure 22. Facility Threat Susceptibility Table

Facility	Resources or Assets Peculiar to the Facility	Likely and Peculiar Hazards and Threats to which the Facility is Susceptible
Refinery	Employees Residents, Records Environment, Building, Equipment, Machinery, Materials	Natural Phenomenon, Sickness, Accidents, Theft, Substance Abuse, Robbery, Contamination, Infestation, Fire, Vandalism, Assault, Kidnapping, Labor Unrest
Retail	Customers, Tenants, Workers, Building, Stores, Records, Cash, Merchandise, Vehicles	Natural Phenomenon, Fire, Vandalism, Riots, Substance Abuse, Sickness Assault, Robbery, Theft, Fraud, Shop Lifting, Pickpockets, Kidnapping
School Campus	Students, Faculty, Building, Records, Equipment, Vehicles	Natural Phenomenon, Vandalism, Riots, Substance Abuse, Sickness Assault, Theft, Robbery, Kidnapping
Shopping Mall	Customers, Tenants, Workers, Building, Stores, Records, Cash, Merchandise, Vehicles	Natural Phenomenon, Assault, Fire Kidnapping, Riots, Robbery, Shop Lifting, Sickness Pickpockets, Fraud Substance Abuse, Theft, Vandalism
Water Reservoir	Employees, Residents, Building, Environment, Records, Materials, Machinery, Equipment	Natural Phenomenon, Sickness, Accidents, Substance Abuse, Infestation, Fire, Vandalism, Robbery, Assault, Kidnapping, Theft, Robbery, Labor Unrest, Contamination

Airline Security on the other hand has terrorism, hijacking, engine failure and extreme weather condition as its major threats and its most critical resources are its passengers and the aircraft.

The security principles upon which the security measures and systems to be used on both facilities are the same. All the aspects of security shall also be integrated and used. But due to the difference in the nature of the threats and the nature of the businesses, an X-ray machine is more appropriate for screening passengers than for bank clients. Radars are applicable to airlines but not to banks.

Another example is the security of an important person (VIP Security) or what is also call Executive Protection. Executive protection does not only require armed escort but rather, it requires all the principles and aspects of security to be applied at his home, office, while in public places or while on travel.

SECURITY SYSTEM DESIGN

Security System Design is basically the process of determining, planning and constructing the most appropriate and applicable combination of security measures, equipment, implements and procedures to address a single or several threats or risks. In most cases of corporate applications, it should not sacrifice expediency, comfort and cost effectiveness.

The basic attributes to consider in designing a Security System are the following:

1. The system must cover as many types of threat or risks as possible with efficiency and effectiveness. Efficiency refers to the system's ability to produce consistent, accurate and timely outputs with the least possible use of human intervention. Effectiveness on the other hand, refers to the capability of the system to produce the desired result at all times.

2. The system must provide measures to address and satisfy all the phases of the security functions cycle which are prevention, reaction and investigation. More weight should be given to prevention. It should be able to deter, detect, delay, deny, diffuse and document a security incident.

3. The system must satisfy the specific principle of security "in depth". It should provide adequate redundancy and at the same time preclude the development of "dependency syndrome" by the operators. It should not unduly inconvenience and intrude into the privacy of the subjects.

4. The system must have provisions for capability for a seamless interface in case of reduction or expansion of infrastructure

and for upgrade to minimize the effect of obsolescence of technology and degradation of material.

5. The system must be user friendly so that its operation can be easily learned, understood, cascaded and that it has wider tolerance for human frailty and faults. For example, an X-ray machine requires a good amount of human skill to detect an explosive because C4 or TNT, both having a lot of carbon in their composition, will look like fruits or cloth in the monitor. On the other hand, an explosive detector using Ion Mobilization System would actually display the name of the substance detected, it hardly uses human skill to detect and identify explosives

6. The system must be cost effective in its acquisition, operation, maintenance and repair. It must have readily available substitute components and parts.

These attributes may be assigned weights to come up with a more systematic way of determining the most suitable system. This is assuming that there are several alternatives or more. An example of this exercise on Anti-intrusion System is as follow:

1.	Preventive Function	60%
2.	Reactive Function	10%
3.	Investigative Function	5%
4.	Redundancy Requirement	5%
5.	Compatibility with other systems	5%
6.	Satisfaction for Seamless Upgrade	5%
7.	Affordable to Mainstream Users	10%
	Total	100%

As was taken up in the first chapter, the purpose of security or a security program is to assist and enable the company in the attainment of its goals.

JOEL JESUS M. SUPAN

The basic mind-set that the organization should maintain is that the effectiveness of security is not measured by how many culprits are caught or by how many crises they have successfully addressed, but by the prevention of any untoward incident so that there would be no culprit to catch nor disaster to address in the first place.

STEPS IN DESIGNING SECURITY SYSTEMS

The following processes are fundamental to the design of a security system.

1. **Determine the hazards that put your team and its resources at the high risk and conduct a Threat Differentiation for each identified threat.** As discussed in Chapter 1, Threat Differentiation is the process of breaking down of the elements of a mishap or a security incident. *(See also Figure 5. Threat Differentiation and the Security Incident Triangle).* The objective of this process is to isolate each of the element or sub-element of the hazard and apply the necessary measure to eliminate the elements. The elimination of any element will eliminate the hazard altogether.

    1.1   The three elements of any security incidents are the following:

        1.1.1   Hazard, threat or risk;

        1.1.2   Object of hazard, target or victim;

        1.1.3   Opportunity.

        A mishap cannot occur if any of the above elements is absent. These elements can further be differentiated by their respective sub-elements.

    1.2   The sub-elements of a **hazard, threat or risk** are:

        1.2.1   **Person or source**. Persons can be any human

being with the faculty of intelligence and emotions. A source can be tangible or intangible. Tangible sources of hazards are animal or physical objects. Intangible sources can be the natural elements such as the wind, the sea, the earth or the sun.

1.2.2   **Capability**. This is the ability and capability of a person to perpetuate his criminal intention. This is also the force or strength and potential of a source to cause damage.

1.2.3   **Motive, drive or tendency**. Motive applies to people who have the faculty of intelligence to discern and reason to perpetrate his intent. The motive can, in most cases, either be greed or need. Animals do not have motive but they are driven by instinct for their survival. On the other hand, the laws of science govern the tendency or potential of inanimate objects or phenomena to become hazards that could cause mishaps.

Thus, hazards or threats cannot exist if it there were no person or a source. It can also not exist if that person did not have the capability or the source did not have the potency. Or, it cannot exist or be eliminated if the capability of the person were removed. As an example, the capability of the terrorists can be removed by cutting off the source of funding. Lastly, the hazards cannot exist if the person did not have motive or the source had no tendency to happen.

1.3.   The sub-elements of the **object of hazard, target or victim** are:

1.3.1 **Susceptibility**. Susceptibility of a target or object of the hazard reinforces the motive of the hazards. A good example is a public place as being susceptible to terrorism because the terrorist is motivated by wide publicity, maximum loss of lives and maximum damage to property to sow terror. On the other hand, a private house is not susceptible to terrorism because it cannot attract public attention. In the same manner, a poor man is not susceptible to kidnapping if the motive of the perpetrator were money. A man is not susceptible to rape because rape as defined by penal codes, applies to women.

1.3.2 **Vulnerability**. Vulnerability as previously defined is the absence of defense against a specific hazard. It is the condition that the perpetrator would take advantage of, relative to his capability.

1.3.3 **Target Asset or Victim**. This is the object that is susceptible and vulnerable to the hazard. It could be a single asset, a facility, a person or a group of persons. The object of the hazard, the victim or target cannot become such if any of the sub-elements is absent and therefore a mishap cannot happen.

1.4 **Opportunity** is the instance in which the time, place and obtaining condition coincide to allow the mishap to happen. Its sub-elements are:

1.4.1 **Time**. Time refers to the instance and duration in which the perpetrator will execute his plan. A perpetrator of a crime knows amount of time he needs to perpetrate his crime. If that time were exhausted, he will most likely abandon his intention. Time is also the exact moment of serendipity when the object of hazard meets the hazard. If that time were eliminated, it will not give the perpetrator the opportunity to carry out his intentions.

1.4.2 **Place**. Place as a sub-element of opportunity refers to where the hazard and its object meet each other for the mishap to occur.

1.4.3 **Obtaining Condition**. This is the surrounding condition that would be favorable for the occurrence of a mishap or not. An obtaining condition may mean the onset of darkness or the isolation of the place from people or its remoteness from the authorities. It can be the obvious apathy of people or team members in a facility. It can also be submissiveness or fearful pre-disposition or panic. These are the conditions that a perpetrator is looking for to decide whether he would pursue his intention or not. A victim who has the presence of mind and is predisposed to evade an assailant can mostly make the assailant abandon his intention. This also applies to a potential target facility whose team members are trained and predisposed to handling such threats.

The absence of any of these sub-elements will eliminate the other elements. The elimination of the any of the elements will eliminate the hazard. The elimination of the hazards will prevent a mishap from occurring.

One event that could serve as an example to this is robbery.

JOEL JESUS M. SUPAN

For robbery to be perpetrated, first, there must be a potential robber. Second, there must be a potential victim and third, there must be an opportunity for the potential robber and the potential victim to meet. If one element were missing, robbery cannot happen.

The potential robber can be further differentiated into three sub-elements. The sub-elements are: first is the person; the second is his capability such as his tools and skills; and the third is his motive. A person may have the capability, but if he did not have the motive to commit robbery, he could not be a robber. Without a robber, there could be no robbery.

Since the basic motive of robbery is to gain a considerable amount of money, if money were removed or reduced, it de-motivates the robber.

The potential target can further be differentiated into three sub-elements. First is the susceptibility. A robber does not randomly pick his targets. He has an intuitive ability to select his target. Curiously enough, there are facilities, which are more susceptible to being targets of robbery than others. The second sub-element is vulnerability. If the target facility appeared to have no adequate security measures in place, the robber would have a soft target. The third sub-element element is the obtaining condition. The robber would always to a preliminary survey of his target before he makes a plan. He would looks for gaps in the security system such as busted lights, sleeping guards, traffic conditions and people in the facility unmindful of their surroundings. This is the obtaining condition that the potential robber is looking for.

Once the elements have been differentiated, it would be easier to determine the kind or mode of security measure or procedure to be applied. As in the case of robbery, sometimes, all it takes is a procedure to remove the motive.

By constantly maintaining a small amount of cash in the till and advertising it, the potential robber can be de-motivated. That would be enough for a robber to look elsewhere and not waste his time and risk being caught for a small amount of money. Even if robbery occurs, the loss would be very minimal and certainly, the robber will not strike again.

2. **Design an Integrated Security System**. Security Integration is the process of using all applicable aspects of security and security systems to complement and supplement each other and address all threat elements individually and collectively. It is a process of providing the widest possible cover with reasonable redundancy and depth to prevent a mishap. It has all the attributes of an ideal security system.

    This concept of security integration can be best illustrated by what was done to a certain gasoline station. The said gasoline station was being held up every so often. This is because, by the nature of the business, the owners cannot put a fence around it lest no client will patronize it. The gasoline station is also along the road and its transactions are mostly in cash. It is both susceptible and vulnerable to robbery.

    At first, to prevent it form being held-up in consideration to the inherent conditions, the owner thought of contracting a security guard team to provide guards for the station. Since a guard is needed twenty-four hours a day, the owner has to have three guards at one guard per shift. He shouldered the cost of the guards , which was equivalent to the prescribed minimum wage plus the mandated benefits.

    Despite the guard, the gas station was held up again. This time, two robbers perpetrated the robbery. The first thing they did was to neutralize the guard.

    In this incident, not only did the station lose all its earnings for the day but also the money, radio and firearm of the guard. It could have been worst if the guard engaged the robber. Casualty could have mostly likely be incurred. In a subsequent study, it was recommended that the guards be removed and it was also recommended that steel drop box with combination lock be acquired, installed and anchored to the floor.

Cameras, public address system and emergency button were installed. A procedure that the cash at the till was maintained at a minimum amount and that in case of a robbery, all the money in the till will be surrendered to the robber. The amount of cash was also advertised. The integrated system was advertised.

Since then, the gas station has not been held-up. But if ever it were held up again, the most that the robber could have taken was the minimal cash that was maintained in the drop box. In all likelihood, the robbers would not come back any more. They cannot risk being caught for a small amount. The gas station in the mean time has not spent on expensive guards anymore.

In the foregoing example, the use of a drop box and the adoption of a procedure to maintain a minimal amount in the till and its advertisement have actually de-motivated the potential robber. The amount of money at the till is the basic motivation of the robber to perpetrate the crime. By minimizing the amount of money and advertising it, a potential robber would be discouraged because the risk he would take in going to prison is much greater than the benefit he could gain from the amount he would get in the heist.

Summary

The design of any Security System should have sufficient breadth to cover as many types of risks as possible and reasonable redundancy for prevention, It must also provide measure to address a mishap and facility to document the mishap for a thorough investigation that can provide basis for the creation of corrective measures to prevent the same mishap from happening again. It should also consider cost of acquisition and its maintenance and reasonable convenience of the users as well as those subject to it.

CHAPTER 13

Emergency Planning, Crisis Management and Business Continuity

Crisis Management is the second phase of the Security Functions Cycle. This is an indispensable component of the security system on account of the principle, "There is no absolute security". This is also another way of stating Murphy's Law — "If anything can go wrong, it will." The ultimate responsibility of contingency planning belongs to the highest level of management. Crisis Management encompasses crisis preparedness, emergency response and crisis contingency.

Emergency is a condition that is developed at the onset of a mishap. It may or may not progress to a crisis condition depending on the magnitude of the mishap and the effectiveness of the response.

While crisis is not always the consequence of a common mishap, it could develop into it if the said mishap were not addressed with timeliness and with the correct emergency response. Crisis comes to be if such resources did not have protection or when all the preventive measures needed to protect them had failed and the impending consequence would be the disruption of the organization's operation. **Crisis** is an unstable and crucial time or a state of event in which an adverse change is impending with the high possibility of an undesirable result. It is a condition of grave danger, extreme uncertainty and it is the turning point towards disaster.

The preparedness program and the contingency plan to the address both emergency and crisis shall be collectively called as Crisis Preparedness Program.

FORMS OF EMERGENCY AND CRISIS

Emergency and Crisis can be a result or consequence of any of the following events but not limited to:

1. Adverse judicial ruling where the organization is involved.
2. Accidents
3. Arrest of Executive
4. Civil Strife
5. Criminal Incident
6. Environmental Threat
7. Extortion
8. Failed major project or undertaking
9. Financial Collapse
10. Fire or Explosion
11. Incapacitation or Death of Key Operator
12. Industrial Unrest
13. Industry-driven Crisis
14. Kidnapping of Executive
15. Regulatory Ban
16. Natural Threat or Disasters
17. Negative Executive Action
18. Product or Service Problem
19. Service or Supply Failure
20. Team member's resistance to a decision
21. Terrorism

CHARACTERISTICS OF A CRISIS

Crisis has the following characteristics:

1. It can happen anytime, anywhere, to anyone for whatever cause;

2. It is bound to happen when you are least prepared;

3. It can happen fast;

4. It can generate a lot of attention from the public;

5. It can cause a breakdown of communication among team members;

6. It can escalate to other untoward consequences such as immobilization, damage to property, unrest, harm or death;

7. It can severely disrupt the operation of the organization;

8. It has the potential to become worse.

The inevitable consequences of mishap if not properly addressed are loss or damage to life and property, damage to the environment, loss of jobs and businesses. Therefore the bottom-line objective of crisis management is to prevent further loss or damage to the organization's properties.

What the organization does in the first 24 hours of a crisis can fundamentally affect the organization's reputation and enhance or hinder its ability to deal with the crisis even in post-crisis situation.

ELEMENTS OF CRISIS MANAGEMENT

There are two basic elements of crisis management.

1. **Crisis Preparedness Program.** This is a proactive undertaking that involves the assigning and commissioning of the organization's resources to be dedicated for the execution of a Crisis Contingency Plan.

2. **Crisis Contingency Plan**. This the preemptive action that provides for the execution of various procedures and utilization of resources allocated to prevent an emergency or crisis.

In the formulation of the Crisis Preparedness Program and the Crisis Contingency Plan, it is important to consider the following principles:

1. No two types of risks are the same.

2. There is a type of action for every type of risk. The more
    you sweat in drills the less damage you incur in disasters.

3. The more prepared you are, the more seamless your Business
    Continuity will be.

ELEMENTS OF A CRISIS PREPAREDNESS PROGRAM

The elements of a Crisis Preparedness Program are as follows:

1. **Statement of Crisis Management Policy**. The Crisis
    Management Policy is a manifestation of management giving
    importance to the organization's need for preparedness
    against any threat to protect the employees and their jobs. It
    should be disseminated to all employees and make clear that
    their security awareness will be an essential component to
    avert emergencies and crises.

2. **Crisis Contingency Plan**. A written Crisis Contingency
    Plan must be prepared for every form of Crisis. A written
    Crisis Contingency Plan is the most essential element of
    Crisis Management and preparedness for any contingency.
    It provides for the policy on the organization of the crisis
    contingency team, facilities for the command center,
    transportations,   communication,   accommodations,
    equipment, sources of funds and all the procedures on how
    every contingency must be mobilized and managed. It is
    essential that this written Contingency Plan is followed and
    carried out to the letter and that a recording or journal of
    its compliance be done in an actual incident. Any deviation
    form the prescribed procedure must be justifiable, defended
    and recorded in the said journal.

3. **High Level of Employee Security Awareness**. Security Awareness is the foundation of an ideal security program. The ultimate objective of a crisis contingency plan is security. This awareness may preclude the organization having to implement the Crisis Contingency Plan at all. However in the event of an emergency, every employee must know by heart their individual role in sounding the alarm for emergency, the initial and correct course of action necessary to address the emergency condition and knowing how to execute their roles for subsequent conditions.

4. **Routine Inspection and Audit of Emergency Facilities**. Routine Inspection of the readiness of the emergency facilities is an integral part of the security preparedness program. A designated team must ensure that this is done religiously every day. Make it a duty of an employee to immediately report obvious deficiencies of any emergency equipment.

5. **Training and Periodic Drills as often as practicable**. Training and Drills are the means to ensure the competency of employees in the execution of the Crisis Contingency Plan. Training will provide the knowledge and the drill will provide competency. Drill will also provide the means to identify the weakness or inadequacies of the plan so that appropriate corrections and adjustments can be made.

ELEMENTS OF A CRISIS CONTINGENCY PLAN

The elements of a Contingency Plan are the same as the elements of a standard security plan. However, a separate Contingency Plan must be made for every form or type of risk. Fire, natural disaster, explosion, accidents resulting to injury, epidemic, disruption of supplies and materials, financial crisis among other risks, require different policies and emergency procedures to be addressed. The reason for this is that every form of risk has peculiar characteristics and creates a different condition. The elements of a contingency plan are as follows:

1. **Situation**. This element provides the information on the prevailing conditions and the rationale that justifies the need for the plan. These conditions vary according to the type of facility and the prevalent risks that confronts it. This element provides for the statement of the following:

    1.1   The description of the specific form of risk with all the obtaining conditions such as the magnitude of risk, the susceptibility and the vulnerability of the organization to this risk and the acceptability of the obtaining condition to the organization. This description is obtained from the Risk Assessment. The Security Assessment and Acceptability Table at Figure 18 provide also the information on the resources that are threatened by the risks.

    1.2   The description of all the available resources and support organizations necessary to assist the company in the execution of the plan. These are the members of the organization, the facilities, the government agencies and other private organizations and facilities cognizant of the risk such as hospital, banks, hotels, law enforcement agencies, media outfits, emergency response units, transport facilities, communication facilities, utilities provider and the like.

2. **Objectives of the Plan**. This element is the statement of the specific objectives of the plan. As previously stated, the bottom-line objective of the crisis contingency plan is to prevent further loss or damage, danger or harm to the organization's members and resources.

3. **Implementation**. This element states the different procedures necessary to address the obtaining conditions. It has the following sub-elements or procedures:

    3.1 **Organization**. This sub-element identifies the leadership and the membership of the Contingency Team and their respective functions. The composition may vary depending on the form of the crisis at hand. The members designated to head these teams and units must have the seniority and the competency peculiar to the function he would be assigned to. A typical Crisis Contingency Team is composed of the following:

        3.1.1 **Team Leader**. He is responsible for the leadership, management and command of the Crisis Contingency Team.

        3.1.2 **Secretariat**. The secretariat is responsible and tasked with manning communications lines with the different operating units of the Team, monitoring the compliance to or deviations from the procedures of the operating units and recording all activities of the team and all information provided by the operating teams. It is also tasked with the reporting to higher management all the obtaining conditions for the duration of the activation of the crisis.

        3.1.3 **Operating Teams**. The Operating Teams are all or any of the following teams as required by the contingency. Some teams may be tasked with concurrent functions as deemed necessary.

            3.1.3.1 **Reaction Team**. This team is responsible and tasked with the assessment of the damage, addressing directly the situation brought about by the mishap and correcting or mitigating it or its impact. In case of fire, this is the Fire Fighting Team and in case of industrial waste spillage, this is the Containment Team.

3.1.3.2 **Rescue Team**. This team is responsible for searching, clearing, and evacuation of all persons from the scene to the designated evacuation area and their accounting and documentation.

3.1.3.3 **Medical Team**. This team is tasked with providing first aid and counseling to all injured persons and casualties. They are also tasked with coordinating with the accredited hospitals

3.1.3.4 **Security Team**. This team is tasked with protecting the exposed properties of the organization from further loss or damage. It is also tasked with cordoning and keeping order the scene of the incident and its surrounding areas.

3.1.3.5 **Transportation Team**. This team is tasked with making available all means of transportation and coordination to transfer people to holding and billeting areas as necessary.

3.1.3.6 **Billeting   Team/Security   Team**. This team provides for activating and coordinating the availability of billeting and holding facilities. This team may also be tasked with providing sustenance and clothing for the members or victims of the crisis.

3.1.4 **Support Teams**. The Support Teams are all or any of the following teams as required by the form of contingency. Some teams may be tasked with concurrent functions as deemed necessary.

3.1.4.1 **Finance Team**. This team is tasked with making available the monies required by the operating teams. It may also be tasked with informing the insurance companies that covers the risks.

3.1.4.2 **Government Liaison Team**. This team is tasked with coordinating for assistance and updating the different government agencies and officials cognizant of the crisis at hand.

3.1.4.3 **Media Desk**. The Media desk is tasked with providing objective, complete and accurate information about the incident to the concerned people and the public. It is also tasked with composing, seeking approval and releasing of information to the public. It is also tasked with the monitoring of media coverage to detect and correct any misinformation about the incident. The media desk should be located far from the Command Center to protect the contingency team from unnecessary distractions.

3.2 **Procedure on what to do and who to inform in cases of detection of an emergency condition.** The basic guidelines to be followed by all teams and every team member of the organization in reacting to an emergency are the Principles of Emergency Action, namely:

3.2.1 Check the area if it is safe;

3.2.2 Conduct primary assessment of the accident to determine of it can be averted, stopped or corrected by the witness or responder;

3.2.3 Call for emergency help;

3.2.4 Conduct a secondary assessment of the condition to be relayed to the Emergency Reaction Team.

3.3 **Statement and description of the different levels of emergency.** The level of emergency provides the guideline on what unit of the crisis contingency group shall be convened and the resources to be utilized. There is no fast rule as to differentiating the conditions between levels. For illustration purposes, the leveling may be done as follows:

3.3.1 **Level 0.** This Level means the mishap would have negligible impact on the organization or on any of its resources. Members of the team directly involved in the incident can address and resolve the incident.

3.3.2 **Level 1.** This means that the mishap would have minimal impact on the organization but requires attention and action of the members of a unit.

3.3.3 **Level 2.** This means that the mishap would have

fairly significant impact to the organization and the Crisis Contingency Team would have to be partially activated.

3.3.4   **Level 3**. This means that the mishap would have a significant impact to the organization and it would require the full activation of the Crisis Contingency. Team.

3.4   **Procedures on the mobilization of the Crisis Contingency Team and all support units and the activation of their utilization**. It also includes the procedure in the activation of the Crisis Command Center.

4.   **Administration and Logistics.** This element provides the definition of each member of the Crisis Contingency Team and their respective functions. It also provides the materials necessary for the teams to carry out functions and the authority on how to acquire, mobilize and utilize them. This includes the provision for a Command Center which must have the following items on hand as prescribed in the Crisis Preparedness Program:

4.1   Audio Visual Equipment (LCD Projectors or Monitors)

4.2   Audio-video recording and camera

4.3   Clocks and timers

4.4   Computers and network lines for word processing and data transcription for recording, documentation printing and dissemination.

4.5   Conference Tables

4.6   Food Supplies

4.7   Maps and Technical (building or machinery) Plans

4.8   Manuals (Crisis Contingency Plan, Technical Manuals)

4.9   Medical Supplies

    4.10 Office Supplies and Forms

    4.11 Radio or Audio-Visual Communication System

    4.12 Slate board with adequate writing implements

    4.13 Telephones with a dedicated lines to the management, operating teams, support teams, media and directories

    4.14 Transistor radio and television to monitor news

5. **Command and Communication**. This element provides for the relationships of the members and units of the Crisis Contingency Team. It provides for substitution and succession procedures. It defines the means, lines, back-up lines and redundancy loops of communications. One of the best practices for substitution and succession procedure is called Incident Command System This system considers certain conditions where the designated unit leaders are not present and that communications lines might not be functional. Thus, the Incident Command System provides that the most senior team members shall assume command of the crisis Contingency Team and its units until such time that the designated head of the Crisis Contingency Team or Units is available and ready to discharge his or her functions.

6. **Appendages**. The appendages to the plan include procedures and the detailed instruction on how to address a specific requirement. Examples of these are the designations of unit call signs, the protocols for communication, diagrams to illustrate a process, contact persons and numbers of support units.

BUSINESS CONTINUITY PLAN

The crisis condition will bring about the disruption or outage of certain resources. The Crisis Contingency Group is commonly composed of the key officers of the company organization and may result to scarcity of management resources.

The organizational provision to address this condition and to ensure the continuous availability of supply of resources, product and/or services is called a Business Continuity Plan.

Business Operation is the combination of all the activities of the teams in the organization using all its resources within a period of time until it achieves it objectives. A normal Business Operation is illustrated in the following figure.

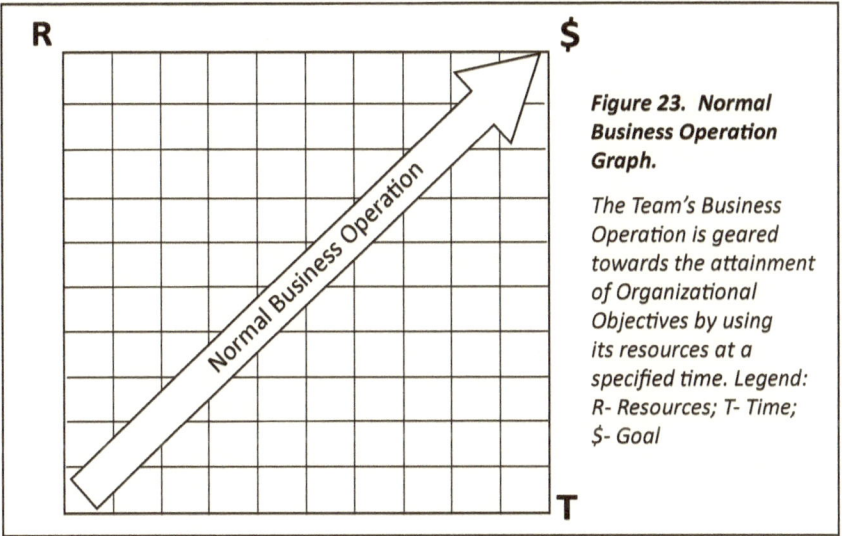

Figure 23. Normal Business Operation Graph.

The Team's Business Operation is geared towards the attainment of Organizational Objectives by using its resources at a specified time. Legend: R- Resources; T- Time; $- Goal

Business Continuity is a concept or undertaking, which prescribes or ensures the availability of resources for the organization to produce and deliver its products or services in the event of a prolonged interruption or disruption of operation as a consequence of a mishap.

A **Business Continuity Plan** on the other hand gives direction, guidance and instructions to the team members individually and collectively as a team on how to address various disruptions of supplies and activities that are occurring in and around the location of the organization's facilities. Its basic objectives are to avoid or minimize the effect of the mishap, to prevent stoppage of business operations, to recover from the disruption and to bring the organization back to its course towards it original objectives.

JOEL JESUS M. SUPAN

It is inevitable to incur diminished resources after a mishap or disruption of business operation, therefore a negative variance from the desired objective will result as illustrated in the following figure.

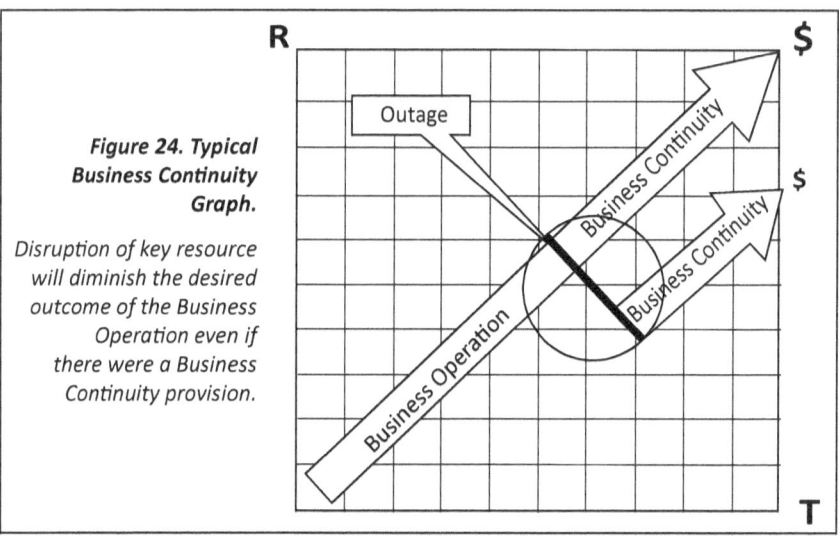

Figure 24. Typical Business Continuity Graph.

Disruption of key resource will diminish the desired outcome of the Business Operation even if there were a Business Continuity provision.

The ideal Business Continuity Plan will provide for extra input of effort and resources so that the organization will be able to recover from the disruption and at the same time achieve its original objective. This is illustrated in the following figure.

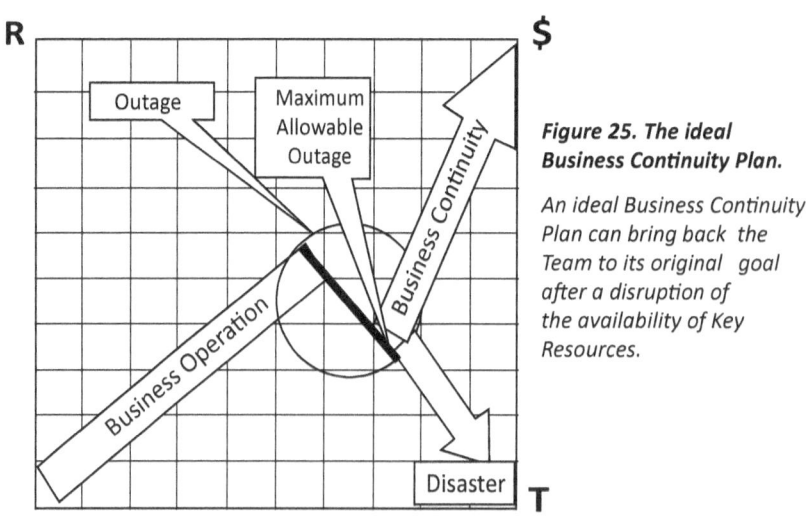

Figure 25. The ideal Business Continuity Plan.

An ideal Business Continuity Plan can bring back the Team to its original goal after a disruption of the availability of Key Resources.

Business Continuity is an integral part of a Crisis Contingency Plan. Considering however, that the activities in the Business Continuity Plan are the same as that of the activities of a normal business operation, it is the means that the organization must have to obtain adequate reserve or alternative sources of materials and replication of information with the least possible cost of inventory, interest or depreciation.

This means that the organization must have a program of identifying and developing members to have multi-skills, documenting and replicating information on company programs and processes and developing and maintaining rapport and goodwill with the sources of needed fund, skills and materials in time of crisis.

Resource Classification	Resources	Expedient Alternative Source
Base Resources	Environment	Sustained Conservation Program, Stand-by Government Permits, Liaison
	Information	Off-site Back Up, Replication, Duplicates
	Finance	Contingency Funds, Credit lines, Financial Institutions
Acquired Resources	Personnel	Intellectual Capital, Succession Plans, Multi-Skills Development, Outsourcing
	Facility	Alternative Site (Owned or Outsourced)
	Equipment	Back-up Inventory, Outsourcing
	Material	Buffer Stock, Alternative Suppliers
Derived Resources (Internal and External)	Product	Generics or Intra-company Sourcing
	Services	Third Party Outsource
	Market	Communication, Public Relations
	Reputation	Communication, Public Relations
Virtual Resource	Time	Availability of all alternative sources.

Figure 26. Table for Expedient Alternative Source of Organizational Resources

The last phase of Crisis Management is extracting lessons from the event and the execution of the Crisis Contingency Plan. In order to extract lessons from the event there has to be a complete analysis of the entire event, the situation or condition before it happened or what caused it to happen, how the teams reacted and the things that the team could have done better. In other words, the entire picture would have to be recreated.

This requirement, thus paves the way for the need to learn investigation.

Summary

Reacting to emergency is the second phase of the functions cycle of security. The only way the losses can be mitigated is to have a timely and a well prepared and executed contingency plan. This can only be realized with the manifest involvement of top management in the preparation of the Crisis Contingency Plan and their sustained and focus drive to the team members to master by heart the plan and their respective tasks. This is the only to avert a crisis situation which can undo what the members made of their team and its good reputation.

-oOo-

CHAPTER 14

Security Investigation and Case

This chapter is intended to give team leaders the basic guidelines on how to conduct an investigation. However, it is not intended to give the detail of investigation procedures of various specific criminal incidents.

Investigation is the primary element of Case Management, which constitutes the third and last phase of the Security Functions cycle. Its primary objective is to find ways and means to prevent the recurrence of a mishap.

Investigation is a planned and organized determination of facts about a specific event, occurrence or a condition for a specific purpose. Security investigation is the process of determining the circumstance and the essential elements of an unusual event that has caused harm, loss or damage to person or properties. It is also the process of determining the factors that have caused the unusual incident to happen. It is initiated when any or some of the elements of an event are not answered in the Incident Report. In it simplest term, investigation is finding the truth.

One of the skills that a team leader should learn is how to conduct an investigation. This is because the team leader, being responsible for the custody and care of the organization's resources, is also responsible for their recovery in case of loss or destruction.

For instance, if a loss incident occurred in an office, the first impulse is call the police or ask the security team to investigate. In all cases, when the police or investigator arrives, the first person they will ask for are the supervisors, the witnesses or the first person in the scene. In the end, the content of the report is actually the statement of those closest to the scene of the incident. However, in most cases, much of the supposed evidence has been erased before the investigator arrived. As a result, most cases remain unsolved and the incident will just be a part of statistics.

Most people believe that investigation is the job of the police or of security. This is so because, no one teaches students that problem-solving, research or thesis is no different from investigation. All of them have the same objectives of drawing the true picture and finding out the truth. All of them involve the gathering of information, determining and establishing their relationships to draw a clear picture of a past event that could lead them to the recovery of a lost or destroyed resource and allow them draw lesson from the incident and find ways to preclude the recurrence of the incident that lead to such loss.

An investigator to be successful needs a broad understanding of different discipline, curiosity, ability to ask the right question and the resourcefulness to gather the answers and a logical mind to determine the relationships between the gathered information. He must also have the facility of the language to be able to compose the report to re-create the true picture that can be understood by those who will read his report.

INVESTIGATION IS A LINE FUNCTION

While investigation is not explicitly stated in anyone's job description other than the person in charge of security, it is in principle, the job as well as the responsibility of the team leader. No one can have a more successful investigation other than the team leaders because they have immediate access to the scene of the crime. They have stock knowledge of the circumstances and they have access to all sources of information.

WHEN TO AND WHY CONDUCT AN INVESTIGATION

An investigation is normally triggered by an incident report that has open-ended questions or requirements that need to be answered. This makes learning how to write an incident report the best way to start learning about security investigation.

An incident report is made when there is a breach in the security chain. This breach could be a potential cause of destruction or loss of resources or harm or injury to the team members. It can also be an event that has actually caused destruction or loss of resources or harm or injury to the team members. Or, it is an observation of destroyed or lost resources or harm or injury to the team members. A good report would pave the way for a successful investigation.

An **Incident Report** is an account of what has been observed. It can be oral or written. For purposes of emphasis and immediate recall of the discussion on reporting in Chapter 8, a good report must be accurate, complete, concise, organized, understandable, neat and timely.

One should remember that an incident report must give, as much as possible, the clearest picture of the incident.

There are instances when the incident report is complete enough that no investigation is necessary. In this case, an incident report may not require an investigation and a subsequent investigation report. This case is called an open and close case.

It is when there is information, which cannot be readily obtained that incident reports cannot give a complete picture. This is when an investigation is necessary.

The ultimate goal of investigation from the perspective of the organization is to find ways and means to prevent the recurrence of the incident with the end in view of pursuing the security requirements of the organization.

JOEL JESUS M. SUPAN

The collective result of all security investigations would enable the organization quantify and determine the cumulative cost of losses resulting from all the security incidents or loss events. This amount can then be used as the baseline for subsequent security programs.

TYPES OF SECURITY INCIDENTS

Incident reports are feedback mechanisms to inform management of the situation within or outside the company. Every team member or leader should know that no incident is too small for management not to know it. It is expected that the team leader who is responsible for the resources of his team should do something to initiate the actions needed to correct a situation.

The three types of unusual incidents are follows:

1. **Abnormal or Unusual Condition**. This situation is just a condition that can potentially lead to destruction or loss of resources or harm or injury to the team members. The witness or observer can readily correct this condition. The observer in this case should readily take action or cause the correction of the condition himself or by informing others who are more capable. But whether the condition is corrected or not, the immediate supervisor and subsequently the higher management should be made aware of the situation. Examples of this are leaking water systems, overloaded electrical systems, blocked exits, unattended machinery, unattended classified document, etc. This unusual incident is the easiest to address as well as it is the easiest to report since all the essential elements are complete.

2. **Unusual Event**. This is a situation where there is an actual incident that resulted to the destruction or loss of resources or the injury of team members. Examples of these incidents are armed robbery or vehicular accidents.

In this case, the reporter is the one who has actually witnessed the incident. Where there are more witnesses, all witnesses are expected to make a report of this incident as they have observed it. It is essential that every witness should make a report so that the incident can be recreated from different vantage points. This can give a clearer picture of the incident for more appropriate courses of preventive action. In case not all of the essential elements of information were obtained, the right questions can be asked as basis for an investigation.

Witnesses generally provide all the essential elements of information except for the cause, the motive or the identity of the perpetrator in case of a crime or a violation of regulation. However, the completeness and accuracy of the report would depend most on how well and how thorough witnesses covered the event and how well they remember what they saw and how well they can write and organize their observation. It is the accuracy and completeness of the report that could lead to the identity of the perpetrator and possibly the determination of the cause or motive and subsequent recovery or compensation for the loss or the correction of the causes of the incident.

Not all witnesses are expected to have the competency to perform proper observation and description, however, a team leader who is competent in observation and description could guide his team or the witnesses on how to completely and accurately describe the event.

3. **Unusual Report**. This is an incident where the result of a past incident is evident. Examples of these incidents are damaged, defaced or missing properties and a team member is harmed. This is a situation where there are no witnesses to the actual and immediate cause for the loss, destruction or harm but it can be inferred that such an event happened

by an obvious and unusual consequence of loss, damage or injury. Examples of these incidents are missing wallet, missing property or death in the work area.

Of all the unusual incidents, this is the most difficult to investigate because there are no witnesses to help the investigator to reconstruct the series of events leading to the effect. All the supposed evidences are mere silent witnesses to the event. This is where the skills of the investigators are most challenged.

But then again, the success of the investigation depends on the investigator. The investigator must also depend on the other means of recording such as logbooks, pictures, video recording and the proper documentation and preservation of the scene of the event. Other tools for investigation, such as interviews and instrumentation for forensics shall be discussed in the forthcoming topics.

OBJECTIVES OF SECURITY INVESTIGATION (As Compared to the Objectives of a Criminal Investigation)

The basic objective of investigation is finding the truth. These basic objectives can be specifically stated as:

1. To reconstruct the circumstances surrounding a security incident or potential security incident;

2. To identify the offenders;

3. To determine who is responsible for the occurrence of the incident or condition;

4. To identify the weaknesses in security that contributed to the incident or unsafe condition;

5. To recover the loss or find restitution;

6. To find ways and means to prevent the recurrence of the incident or unsafe condition;

A **Criminal Investigation** is focused basically of the determination on the elements of a specific crime. Locate the suspect, gather evidence and prove the relationship between the suspect and the crime through the evidence gathered. It does not give regard to the damages incurred by the victim nor to the prevention of the recurrence of the same crime.

Security Investigation on the other hand considers other requirements as learning lessons from the conclusions and finding ways and means to prevent the same incident from happening.

SYSTEMATIC APPROACH TO INVESTIGATION

A systematic approach is necessary for an investigator to have a successful investigation. The following are the steps to have a systematic approach to investigation:

1. **Determine the Investigative Problem**. The investigative problems are those questions that need to be answered in the investigation. The answers to the investigative problem shall compose the true picture of the incident.

2. **Determine the Essential Elements of Information**. The Essential Elements of Information (EEI) provides the framework of an investigation. They are the answers the following questions:

   2.1   What happened?
   2.2   Where did it happen?
   2.3   When did it happen?
   2.4   Who caused it to happen?
   2.5   How did the perpetrator do it?
   2.6   Why did the perpetrator do what he did?

   In a Criminal Incident, the Essential Elements of Information (EEI) will be sufficient for the law enforcer to take action to arrest and persecute the perpetrator. But in most cases, one or several of the questions need to be answered, and unless all the EEIs are determined, no case can progress. To arrive at those EEIs, other questions would have to be raised. Those questions are called other investigative requirements or OIRs. The Security Incident Triangle illustrated in Figure 5 in Chapter 5, is a good instrument to determine which element or factor of the incident is present or absent in the incident and identify the essential elements of information.

3. **Determine the Other Investigative Requirements**. Other Investigative Requirements (OIR) are the answers to other questions that have to be asked to lead the investigators to the essential elements of information. For example, if the identity of the perpetrator is not known, the following questions would have to be asked and the answers have to be determined:

   3.1 How does the perpetrator look like?
   3.2 What are his outstanding characteristics?
   3.3 What are his general characteristics?
   3.4 What did he use?
   3.5 Who were with him?
   3.6 Where did he go?
   3.7 How long did it take him to do what he did?
   3.8 What law was violated?
   3.9 Who among the witnesses knows him?

The OIRs also provide the "flesh" to the EEIs to create a more vivid picture of the incident. More OIRs can be identified depending on how many EEIs are left undetermined.

4. **Identify the reason or need for the Investigation**. The rationale for criminal investigation is to enforce the law, to protect the law abiding citizens, to put order to society by isolating the perpetrator and make such isolation an example for others not to emulate him, to give justice to the victim and

to correct the conduct of the perpetrator. These rationales of criminal investigation, while explicitly stated in the law are generally implied in the course of the specific criminal investigation.

The rationale for security investigation is based on the basic objectives and interests of an organization. Thus, the reason for investigating an accident is to prevent it from happening again because it is the interest of the company to have a safe work environment. In case of loss or destruction of company resources the reason for investigating it is because it is the basic objective of the company to be profitable and to grow. Obviously, the loss of resources is detrimental to the profitability and the growth of the company.

5. **Determine the Objectives of the Investigation.** The objectives of the investigation are basically what the investigator wants to determine. However, as previously discussed, there are fundamental differences between Criminal Investigation and Security Investigation.

The basic objectives of any criminal investigation are generally implied and not explicit. These are as follows:

5.1 To determine if all the elements of the specific crime are present and establish a probable cause

5.2 To identify, locate and apprehend the perpetrator

5.3 To gather the evidence

5.4 To establish the link between the perpetrator to the crime through the evidence.

The number of criminal incidents is finite and all of them are well defined by law. Crime patterns have been established and various tools to gather evidences have been developed. Advances in technology have also made investigation a lot easier for the law enforcers to do and be successful. Thus,

JOEL JESUS M. SUPAN

one will seldom find those objectives to be explicitly stated in a police investigation report.

On the other hand, the causes and reason for loss or destruction of the organization's resources and the causes of harm or injury to the members are infinite. This equally makes an infinite number of ways by which security investigation can be approached. This is compounded by the unavailability of technology to assist in security investigation.

Thus, the objectives of a security investigation have to be stated explicitly and specifically. The success of the investigation will depend on what essential elements of information are obtained.

The basic objective of a security investigation is to gather essential elements of information not obtained and stated in the incident report. For example, if the case is an observed loss of property, the objective should be, "To determine the perpetrator." If the incident is an accident, the objective should be, "To determine the cause of the accident."

The other objectives of a security investigation in consonance with the interests and objective of the company and which are not considered in a criminal investigation are as follows:

5.5   To determine the veracity of the incident report

5.6   To recover or find restitution for the loss

5.7   To find out who is liable or responsible

5.8   To find ways and means to prevent the recurrence of the incident

6. **Collect Relevant Data**. After identifying the investigative requirements. The investigator is now ready to collect the necessary data to determine the EEIs and the OIRs.

To do this, the investigator must plan his courses of action to be able to cover all the objectives at the fastest time possible with the least resources. The plan shall include the following:

6.1   **Resources   and facilities needed to gather the necessary information to answer the said objectives.** These resources could be manpower, the time needed, the logistical requirement or the necessary experts who could provide insights in the course of the investigation; examples are forensic scientists or other investigators who may have encountered the same incident before. Sometimes the requirement could only be realized during the course of the investigation.

6.2   **Sources of Information.** Knowing where to look for the information is said to be half of the investigative effort.   There are basically three ways by which information can be obtained. They are also known as the three I's of Investigation. They are Information, Interview (Interrogation) and Instrumentation.

   6.2.1   **Information.** This refers to the available written information that the investigator can use. This can be obtained from the original report. As part of the objective of security investigation, it is essential to establish the veracity of the report. Basic sources of information are policies and procedures manuals, the code of conduct, operational reports, accomplishment reports, plans, publications or newspaper. The typical sources of information are incident reports, logbooks, journals, reports, manuals, rules and regulations manuals, newspapers, public documents and case files. It is necessary that these sources be exploited before attempting to interview or interrogate any one. The pieces of information obtained from open sources are the best basis for formulating question that can be asked during interviews. They are the most effective means of drawing out answers from interviews. An interviewee would find it difficult to avoid answering or evade the questions if he realized that the investigators has a lot of background information.

Sometimes however, with the right questions being asked, the available information would be sufficient to draw a clear picture of the incident, thus saving a lot of time for the investigator to complete his investigation for a timely report and resolution.

6.2.2 **Interview or Interrogation**. The term interview is basically used for the exchange of ideas between an investigator and a cooperative witness. Interrogation on the other hand, is generally used on a hostile suspect or an uncooperative witness. Here, only the investigator can ask questions and the witness is only expected to answer. For this method to be productive it is important for the investigator to write all the essential questions that need to be answered but he should be quick enough to make a follow through question. This is where his vast knowledge obtained from open sources of information and his stock knowledge on various subject matters can be put to use. He should also be capable enough to know signs of deception.

The exchange of questions and answer can be

written down and documented. This is then called a written statement, which if subscribed to, becomes a sworn statement. The interviewer can also allow the interviewee a free hand in making a spontaneous narration of what he knows about an incident. Upon subscription by the interviewee, this may well serve also as a sworn statement.

A confession or admission of guilt by a perpetrator is the single most potent evidence that can lead to a successful investigation. A single confession that is consistent with the circumstances of an incident can be more than the weight of a hundred pieces of circumstantial evidence.

This method of obtaining information can be very tedious and time consuming especially when there are a lot of potential witnesses, both cooperative and uncooperative.

One method that has been extensively explored and being put to use is the Incident Survey Analysis. This method is based on the principle that a deceptive interviewee can be determined with the way he writes and composes his statement of a particular incident. While this cannot effectively pinpoint the perpetrator, it can eliminate the number of interviewee and narrow down the suspects and save a lot of time. A good investigator can actually draw a confession from the results of this method.

The typical resource persons for interview are the following:

6.2.2.1 Witnesses

6.2.2.2 Friends

6.2.2.3 Enemies

6.2.2.4 Victim/s

6.2.2.5 Professionals

6.2.2.6 Suspect/s

6.2.3 **Instrumentation.** This is a method of extracting information using scientific methods involving physical, chemical and forensic sciences, instruments and technology. The pieces of information obtained using this method are said to be irrefutable and are accepted as factual and therefore accepted as proof to establish the link between the perpetrator and the incident.

Examples of common instruments and processes to extract information are as follows:

6.2.3.1 **Polygraph**. This is a set of machines used to detect and document signs of deception in the responses of a person being interviewed through abnormal changes in the blood pressure, breathing, pulse, acidity of secretions and shrinking of the skin.

6.2.3.2 **Dactyloscopy.** This is a process of lifting fingerprints in a crime scene and comparing it with the fingerprints of a suspect and with those in the data bank.

6.2.3.3 **Forensics**. This the process of gathering and determining the physical and chemical composition, characteristics and sources of trace evidence found in the scene of the crime and on the suspects.

6.2.3.4 **Photography**. This is the process of preserving a scene with the use of a camera either on film or digitally.

6.2.3.5 **Ballistics**. This the process of comparing the physical characteristics of explosive devices, fire arms and their ammunitions, either used or unused, and the chemical composition of the explosives used to determine their sources and relationship with the suspects

6.2.3.6 **Biometric Comparison**. This is the process of comparing the peculiar physical characteristics to that of an existing database of biometric files. They include the face, voice, iris and vein patterns.

7. **Organize the Investigation Team and distribute tasking**. The investigator should know the limitations of his capability. He has to utilize his team in gathering information. He must ensure that he knows the capability of each member of the investigation team.

8. **Determine the identity and personality of the report originator**. This process is essential in establishing the case. The actual witness to the event is in the best position and is the most credible person to make the report. A third party report, such as the report of an investigator or a person in authority, has to specifically indicate the names and roles and whereabouts of all the sources of the information upon

which they have based their report.

9. **Verify and analyze information gathered**. The information must have the following attributes and descriptors:

    9.1 **Factual**. The information gathered must be from first hand observation and witnesses and from original or authenticated documents.

    9.2 **Complete**. The pieces of information must be adequate to provide answers for the objectives and questions set by the investigation.

    9.3 **Appropriate**. The information gathered must be cognizant of the cases and issues at hand.

    9.4 **Verification of Method**. The methods used to gather information must be certified and in accordance with the prescribed procedure or accepted best practices

    9.5 **Documentation of Information**. The pieces of information must be property documented with timeliness and that they were gathered according to prescribed procedures.

    9.6 **Authenticity of Document**. The sources of written documents must be authenticated, certified and issued by competent authorities.

    9.7 **Chain of Custody**. There must be an irrefutable proof or documentation that the chain of custody of the evidence was unbroken.

10. **Draw the Conclusion**. The conclusion is derived from the establishment of the relationships of the information gathered. The conclusions are simply the answers to the objectives set forth by the investigator.

11. **Formulate Recommendations**. Recommendations are the

proposed courses of action to be taken to address and correct the condition that caused the incident to occur, recover the loss, provide restitution for the damage and/or prevent the recurrence of the incident. The recommendations are based on the conclusions arrived at by the investigation in which the causes of the incident are established.

12. **Case Disposition.** This step calls for either the closure of the case or the establishment of its status if no conclusions were arrived at or there were no answers to the set objectives.

PARTS OF A TYPICAL INVESTIGATION REPORT

One of the most essential products of an investigation is the Investigation Report. It is the proof that an investigation has been conducted and it is the basis for future action and disposition of its recipient. As in the conduct of the investigation, the composition of the Investigation Report must be systematic. It must also posses all the qualities of a good incident report. The parts and the description of the contents of the parts are as follows:

1. **Heading.** The heading is where the standard letter or memo headings, date of report, addressee, originator and the subject matter are stated.

2. **Authority.** Stated here are the nature and source of authority being vested on the investigator to have jurisdiction on the case at hand. It may be given orally, in writing or by its being inherent to one's function. Dates and place of issuance must also be stated for reference.

3. **Matter Investigated.** Stated here is the summary of the original report. Stated here are also the reasons, the importance and/or the rationale of the investigation. The

reason or rationale of the investigation basically refers to the interests of the organization that have been challenged by the security incident or mishap. For example, an accident must be investigated because a safe and security environment for the team members is one of the primary interests of the company.

4. **Objectives**. Stated here are pieces of information that the investigation wants to find out, to complete the picture that could reconstruct the incident or condition. Said items of information are the one missing in the report. State here also the standard objectives of all security investigation. They are the determination of the veracity of the incident and find out ways and means to prevent the recurrence of the incident.

5. **Findings**. Stated here are all the investigative activities and the findings and data, as a result of such activities.

6. **Discussions**. Stated here are the analyses of all pieces of information gathered and their relationships with each other to reconstruct the incident. Opinions based on sound assumptions, theories and logic may be stated here.

7. **Conclusions**. Stated here are the conclusions, which are the answers to the set objectives. They should follow the same sequence as that of the objectives.

8. **Recommendations**. Stated here are the courses of action to take to prevent the incident from happening again. They should follow the same sequence of the conclusion.

9. **Disposition**. Stated here are the recommended courses of actions on what status the case will be put in. Stated here also is the status of the case as of the writing of the report. If the investigation has not arrived at a conclusion, the entire report would basically be called a progress report. There

after, the investigation if it were allowed to be continued would have as objectives, the unanswered objectives. When all the objectives were met, then the investigator would recommend for the closing of the case or his relief from the case.

10. **Annexes**. Listed here are the related materials to support issues, statements, and findings. They may include the original incident report, affidavits, authority to conduct investigation, extracts of record books, extracts of rules and regulations, laboratory reports, receipts, certification and public documents.

INCIDENT RECORDING SYSTEM

The organization must adopt a recording system for all security incidents. This record will provide statistics of all the types of incidents, as well as the actual cost and incidental losses. The total loss can be estimated by using the formula, similar to the formula used in the assessment of the criticality of an asset, adopted as follows:

$$TL = (Lr + Ls + Lo + Li + Le) - (I - P) - R$$

Where:

TL - Total Loss

Lr - Cost of replacement

Ls - Cost of temporary substitute

Lo - Other Related cost

Li - Cost of Income loss

Le^* - Cost of executive time for Investigative activities

I - Indemnity by insurance or perpetrator

P - Amount of Insurance Premium

R - Restitution

Cost of executive time (Le) is computed by adding the total amount of time all persons involved actively in the investigation and disposition of the case multiplied by their respective hourly salaries.

The summation of the costs of all mishaps and incidents can be used as the baseline figure for the losses of the company for the past year. This shall now be used as the reference for the future security objective of the organization. That objective is the reduction, or the elimination of loss.

Summary

Investigation is the last phase of the functions cycle of security. The security program cannot be complete without the adoption of a policy on the investigation of security incidents. Investigation is a competency which every leader and every team member must acquire. It must be accepted that once a security incident happened, loss has been incurred. More losses can be incurred if the investigation were not done with timeliness on account of the loss of productive time by all those concerned in and with the incident. The losses can be further compounded by the possibility of the recurrence of the incident because no solutions were adopted due to the failure of the investigation.

-oOo-

Afterword

THE STONEWALL SECURITY MODEL

The manner by which the topics were presented in this book follows the phases of the functional cycle of security namely, prevention, reaction and investigation. The presentation of the aspects of security also, to some extent, followed the classifications of resources namely, base resources, acquired resources, derived resources and virtual resources. The sequences are based on a security model, which I have called The Stonewall Security Model. The Stonewall Security Model was developed to illustrate an integrated system using a seamless interface of the six aspects of Security.

Figure 27. The Stonewall Security Model. *The Stonewall Security Model illustrates an integrated system using a seamless interface of the six aspects of Security.*

The integration of the six basic aspects of security represents a chain symbolizing security. Each aspect represents a link; the six links are formed into an unbroken circular chain.

At the center of the chain are the team's resources, which are the objects of security. Each aspect is composed of elements, which are the security measures by which the aspect is applied.

The assets are surrounded by three sectors represented by arrows going around the resources. They represent the three phases of the security functions cycle, which are prevention, reaction and investigation. Every element or aspect of the security chain must consider the three phases of security function. Each aspect and element must inherently be a means of protection or the preventive factor for a specific hazard. Prevention is the basic essence of security.

But since there is no absolute security, some mishaps could not be avoided or prevented. Hence, contingency procedures or reactive measures must be adopted to respond to mishaps to prevent further loss or damage.

The aspect or element must also have systems provision for investigation to recover the loss and to prevent the recurrence of the incident.

Should any of the aspects or any of its elements be missing, the resulting gap will constitute the weakest link in the security chain; it will render the entire chain useless for protection. It is at this gap where the hazardous incident or mishap will certainly come from. This represents the second security principle, which states that security is only as good as its weakest link.

The interlinking of the aspects, the elements and the security functions cycle and the supplementary and complimentary effect they have on each other represents the third basic principle of security, which states that high relative security can be achieved in depth.

The resources at the middle of the chain also represent the different functional units of the team. The teams are composed of the team leader and the members. These resources represent the fourth basic principle of security, which states that security is everybody's concern.

The security function cycle by itself represents another principle, which is, "no two installations are alike." All facilities vary from one another, the differences in circumstances mean the assets being surrounded can be exposed to different kinds and/or degrees of threats or dangers; thus, they require different types and/or degrees of protection.

Stonewall symbolizes what is strong, relatively impenetrable and lasting. It also symbolizes what is basic. Stone is the most basic material for fortifications to protect camps and forts against the enemies. It may well also be the best weapon to drive the enemies away. It is impervious to the elements. Rain, wind or fire can hardly destroy it. It is a tool for survival. Since its existence, man has been using the stone for survival. It withstood the test of time. It was with stones that the pyramids, the Great Wall of China and the Stonehenge were built. Its abundance makes it so common that it is the least expensive material for fortification. A stone symbolizes what is simple, useful, strong and lasting. A wall is a product of man's intelligence, creativity and diligence. Stonewall is therefore a combination of the resources provided by nature and the resource provided my man. It aptly symbolizes security.

The Stonewall Security Model provides what is basic, effective and cost effective. It educates the traditional security practitioners, the consumers of security services and team leaders on what security is all about, its basic principles, the aspects which are covered by the principles and the different systems, modalities, procedures and elements by which the said aspects are converged, integrated and applied.

JOEL JESUS M. SUPAN

The Stonewall Security Model shall provide the fundamentals of security integration where every principle is applied by making all the aspects complimentary to the other and supplementary to the whole.

Practitioners can use the Stonewall Security Model in a lot of ways. It can serve as the checklist in conducting a security survey. It provides a systematic guide for the surveyor to follow to be thorough in his survey. It is a good training tool as it sets the learning standard for all those involved in the management and operations of the team.

It can also be used as a model in developing a security plan and program. It illustrates the roles and areas of concern of every team leader.

It can be used to illustrate a security assessment by showing the vulnerability and the extent of such in degrees in relation to the obtaining present security set up of an Organization. Figures 28 and 29 are examples of illustrating a security assessment using the Security Model.

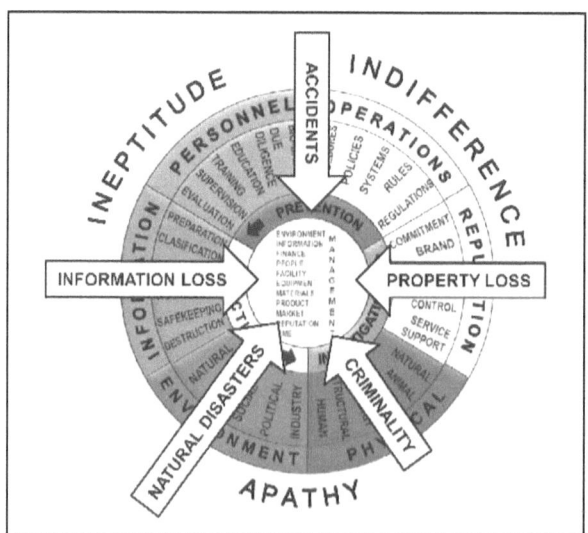

Figure 28. Illustrations of Vulnerability Assessment. *An Organization with inappropriate and inadequate security and compounded by inept, indifferent and disloyal members is vulnerable to mishaps that can impede the achievement of its goals.*

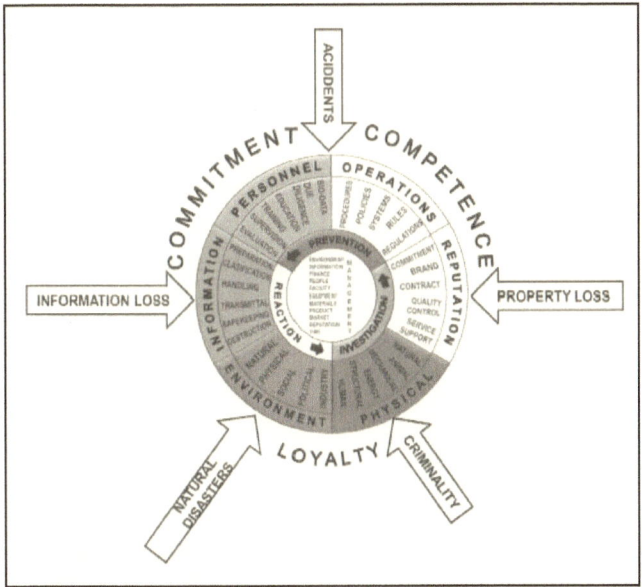

Figure 29. Stonewall Security Model illustrating the resiliency of an Organization. An Organization with committed, competent and loyal team members with the appropriate and adequate security is sure of achieving its goals.

It should be noted that every security requirement that needs to be addressed necessitates the application of all the aspects of security.

Taking the same example in Chapter 1 on Information Technology (IT) Security. The employee as a hazard to information security requires the application of personnel security. The same is true with physical security and all other breaches of security measures. Their application is necessary for all other threats such as intrusion to the server room.

Another example is that of executive protection. The conventional mind-set for executive protection is the training of armed escorts and drivers. In the context of this model, the escorts and drivers are just the guards, which is one of the three types of human barrier. The human barrier is just one of the six aspects of security. Executive protection always starts with the members of the household and with the question, "Are the members of the executive household capable, reliable, trustworthy and loyal?" This is nothing less than Personnel Security. All the other aspects of security are needed to provide for the entire operation of executive protection.

The model can be applied to the smallest unit of enterprise, which is the person himself. It can also be applied to the next larger unit called the family as much as it can be applied in every unit of the team and the whole team, organization, community or even the country, region and mother earth.

Summary

The Stonewall Security Model can best illustrate what security is all about, that is "Security in its broadest sense is a relatively predictable environment where one can pursue its objectives of efficiency, stability, profitability, growth and Sustainability and promote such interests without fear from the occurrence of the hazards that threaten it.

All the Security aspects and elements described in this book revolve around the Stonewall Security Model.

-oOo-

ACKNOWLEDGEMENT

One of the challenges I have encountered in writing this book is the absence of references for the terms that would constitute and integrate all the aspects and elements into a single body of knowledge. Most of the concepts presented in this book were based on or developed from the handouts from the different courses offered at the Naval Intelligence Training Institute of the Philippine Navy and the Philippine Society for Industrial Security. As in most schools and seminars that I have attended, their handouts do not have citations as to the source, thus making it difficult for me to cite or acknowledge specific source or particular person. I have also come to realize that creation of new ideas and concepts can be done by several individuals in different places at the same time, without them knowing of the existence of the other. With the trend of simplifying the manner by which ideas are to be expressed, it is very likely that the same concepts can be expressed in exactly the same way by different authors in different places at the same time. This also accounts for the seeming absence of references cited in this book.

Nevertheless, I would like to cite the Webster Dictionary and the Wikipedia as the primary sources in defining various terms that I have used.

But it is my math professor, Ronald Mendoza, whom I would like to thank for giving me the idea on how to appropriately define terms. He has imparted to me that in Mathematics, regardless of what symbol or term is used in a given equation it is fundamental to define such symbol or term to be valid.

Chapter 8 of this book was entirely extracted from the book "The Art and Science of Guarding" that I wrote in 2002.

JOEL JESUS M. SUPAN

I would like to also acknowledge my company for providing me the space and opportunity to validate the value of the concepts presented in this book.

I would like to thank my team leaders and peers, Gerard Aquino, Aldo Barrios, Ava Engel, Leo Gonzalez and Sharyn Varron-Jacobsen in particular and all the team members of my company in general, who all have tacitly given their affirmations to the concepts by putting them to use.

I would like also to thank my wife Susan and my sons Martin Thomas, Stephen Isaac and Jan Nathaniel for providing me the inspiration, space and encouragement to write this book.

Lastly, I would like to dedicate this book to all the Team Leaders whose primary responsibility is to protect the resources of their respective teams and to all security practitioners who have a frontier ahead of them in the pursuit of their chosen profession.

-oOo-

ABOUT THE AUTHOR

The author, JOEL JESUS M. SUPAN, started his learning on security and guarding as a Cadet of the Philippine Military Academy. After graduation, he became an Officer of the Philippine Navy. After his shipboard assignments, he joined the Naval Intelligence where he eventually became the Course Director for the Naval Officer's Intelligence Course. Thereafter, he practiced his security profession as a Director for Security of a high-end hotel and as a General Manager of a noted guard force provider. He also became the Senior Consultant and Head of Education and Training of a noted risk control and security consulting firm. He has done successful surveys for the biggest petroleum companies, upscale shopping malls, modern hospitals, one of the biggest car manufacturers, the Clark Airbase which was once the biggest U.S. Military Base outside the continental USA. He has been accepted as member of the Philippine Society for Industrial Security and the American Society for Industrial Security. He retired as Vice-President for Security and Compliance of the largest shipping line in the Philippines. He has been appointed by the Commission on Higher Education of the Philippines as Chairman of the Technical Committee that formulated the curriculum for the Bachelor of Science in Industrial Security Administration. He now lives in the Manila, Philippines with his wife and three sons whole all share in the belief that,

"Security is the bridge to every goal...to every dream."

www.ingramcontent.com/pod-product-compliance
Lightning Source LLC
Chambersburg PA
CBHW032058280526
45784CB00012B/36